Applications in Operational Culture

Perspectives from the Field

MARINE CORPS UNIVERSITY
QUANTICO, VIRGINIA
2009

Edited by
Paula Holmes-Eber
Patrice M. Scanlon
Andrea L. Hamlen
Marine Corps University

Published by Books Express Publishing
Copyright © Books Express, 2011
ISBN 978-1780390-32-1
To obtain further copies of this book please contact
info@books-express.com

Contents

Foreword ix
By Major General Donald R. Gardner, USMC (Ret)

Acknowledgments xi

Introduction 1
Operational Culture and the Marine Corps: A Historical
Perspective 3
    Creating Sound Cultural Analyses for the Military 5
    Lessons Learned 7
    Planning for the 21st Century 10

Chapter One: Maslow is Non-Deployable: Modifying Maslow's
Hierarchy for Contemporary Counterinsurgency
    by Major Jonathan Dunne, USMC 15
    Statement of the Problem 16
    Operational Culture Defined 16
    Understanding Ourselves through Maslow 19
    The Iraq Case: Reassessing Maslow 19
    Discussion of Findings 23
    Conclusion 26

Chapter Two: The Use of Cultural Studies in Military Operations: A
Model for Assessing Values-Based Differences
    by Lieutenant Colonel Alejandro P. Briceno, USMC 29
    Culture and Military Operations 31
    Cultural Values Defined 32
    Military Application of Cultural Intelligence 35
    A Cultural Values Model 37
    Applications of the Model to a Case Study: Kuwait 40
    Discussion of Findings 43
    Conclusion 44

Chapter Three: Developing the Iraqi Army: The Long Fight in the
Long War
    by Major John F. Bilas, USMC 49

Methodology 50
Background Information 52
The Surge—Operation Fardh-Al Qanoon
and the Awakening..55
How the Surge Affected Logistics, Personnel, and Pay 59
Iraqi Brigade (1st BDE, 2d Div) Progress in Mosul, Iraq 68
Discussion 69
Recommendations 70
Conclusion 74

Chapter Four: The Way Ahead: Reclaiming the Pashtun Tribes through Joint Tribal Engagement
by Major Randall S. Hoffman, USMC 79
Afghanistan at the Crossroads 82
Tribal Engagement: The Center of Gravity 83
The Pashtun: A House Divided..85
Reengaging the Pashtuns to Strengthen
Kabul's Legitimacy 98
Breakdown and Relationship of the Joint Teams 101
Conclusion 113

Chapter Five: The Application of Cultural Military Education for 2025
by Major Robert T. Castro, USMC............................117
Cultural Anthropology and Irregular Warfare Operations......118
Getting Culture on Board..125
The Reinvention of the Wheel: Culture in Past
Military Operations 126
Educating for the Future 128
Conclusion 132

Chapter Six: Operational Culture: Is the Australian Army Driving the Train or Left Standing at the Station?
By Lieutenant Colonel Steven Brain, Australian Army 135
Defining Culture 137
Culture and Future Warfare 141
The Australian Army: Operational Culture in Education
and Training...143
Conclusion 151

Conclusion 155
   Cultural Role of Time 156
   Military Personnel........ 156
   Cultural Reality of the Area of Operations 158
   Cultural Challenges of our Own Military 159

Appendixes
   Appendix A. Iraqi Army Logistics Data, 2006-2007 161
   Appendix B. Iraqi Army Pay, 2006-2007 165
   Appendix C. Growth of the Iraqi Army 171
   Appendix D. Case Examples of Culture and
   Operations Iraq 173

List of Illustrations

Figure 1: CAOCL's Operational Culture Construct 17
Figure 2: Individual Well-being and Achievement 18
Figure 3: Operational COIN Paradigm and the Cultural Barrier: OIF 20
Figure 4: Regional and Local Growth & Prosperity: Operational Culture Construct: OIF 22
Figure 5: Modifying Doctrinal IPB to Incorporate Operational Culture 24
Figure 6: Hypothetical Cultural Values Model 36
Figure 7: Hypothetical Cultural Values Model using Religion as Example 39
Figure 8: Kuwaiti Cultural Values Baseline 41
Figure 9: Kuwaiti Cultural Values Baseline: National ID 42
Figure 10: Effects of Operation Fardh Al-Qanoon on Iraqi Provinces 53
Figure 11: Iraqi Army Operational Readiness Assessment (ORA) Level Definitions 58
Figure 12: Personnel Strength for 2nd Brigade, 7th Iraqi Division (2006-2007) 63
Figure 13: Personnel Strength In Comparison To MTOE (2006-2007) 63
Figure 14: Pashtuns: The Largest Ethnic Group in Afghanistan 80
Figure 15: Pakistan's Federally Administered Tribal Areas 81
Figure 16: Khost Province Tribal Map 92
Figure 17: Breakdown and Relationship of the Joint Teams 100
Figure 18: Example of a Joint U.S.-Pakistani Command Structure to Support Pakistani JTTs 110
Figure 19: Continuum–Framework for All Things Training and Education 129
Figure 20: Officer Career Progression of Operational Culture, Professional Military Education 130
Figure 21: Enlisted Career Progression of Operational Culture, Professional Military Training 131
Figure 22: Levels of Operational Culture 138
Figure 23: Cultural Hierarchy 140

Figure 24: Standard Peacetime Training Objectives Applied against the Cultural Hierarchy 145
Figure 25: Standard Operational Deployment Objectives Applied against the Cultural Hierarchy 147
Figure 26: The Human Systems Company 151

Foreword

On behalf of the Marine Corps University, I am proud to present this collection of six outstanding essays on operational culture by the officers at Command and Staff and College and the School of Advanced Warfighting. At MCU, our mission is to prepare our Marines and fellow service members to meet the challenges of future operating environments. The officers who attend our program come not only to learn, but also to share their operational experiences and knowledge with each other, advancing our understanding of the new and changing warfare environments where they deploy.

The Marine Corps is first and foremost an expeditionary force, operating in foreign environments among foreign peoples. In order to achieve their mission, Marines must be always ready to adapt to new and challenging conditions. Today, as the U.S. military prepares for warfare in the 21st Century, our ability to think critically and creatively in order to find new solutions to complex situations becomes even more significant. This demands not only traditional warfighting skills, but also new skills in counterinsurgency, stability operations, working with foreign military and security forces, and understanding the local population. Central to these skills is the ability to understand and work within foreign cultures.

The following pages contain six chapters discussing the importance of incorporating culture into current and future operations. Several of the authors draw extensively on their own personal operating experiences in Iraq and Afghanistan, providing important "boots on the ground" perspectives. Others carefully evaluate our current capacity in light of the cultural challenges we will face in future deployments around the globe: offering potential solutions for the future.

These chapters offer an opportunity to capture and learn important cultural lessons from our Marines and fellow service members as they pass through the halls of the university, lessons that I hope will be learned and passed on to the succeeding generations of officers who not only attend the university, but train and prepare for operations around the globe in the 21st century.

DONALD R. GARDNER
Major General, U.S. Marine Corps (Retired)
President Emeritus, Marine Corps University

Acknowledgments

The authors of this book would like to acknowledge the support and contributions of a number of individuals who have made this publication possible. First we would like to thank the faculty and staff at the Marine Corps University (MCU) and the Marine Corps University Press for their faith in this project and their involvement, whether through mentoring masters theses, reading drafts of the text, or providing funding and staff to assist in the process. In particular, we would like to thank Major General Donald R. Gardner (Ret) for his vision and support, not only of this book, but for the development of a robust and integrated cultural curriculum across the PME spectrum. Without this vision, neither the impetus for this book nor the enthusiastic participation of the students and faculty in writing and mentoring these papers would have been possible.

Additionally, a number of faculty members at Marine Corps University spent a considerable amount of time reading and critiquing the various chapters and we would like to extend our thanks to Dr. Wray R. Johnson, Dr. Amin Tarzi, Lieutenant Colonel Alex Vohr, Dr. Paul D. Gelpi, and Professor Erin Simpson. Dr. Jerre Wilson and the staff of the Vice President of Academic Affairs were central in supporting and funding this work. In addition, Dr. Charles P. Neimeyer, director; Kenneth H. Williams, senior editor; W. Stephen Hill; Wanda J. Renfrow; Emily D. Funderburke; and the staff at MCU Press were, as always, critical to the production and preparation of the manuscript for publication.

For reading drafts, commenting on the material, and numerous discussions on the challenges of applying culture to military operations, we would like to thank Colonel George M. Dallas, USMC (Ret), director of the Center for Advanced Operational Culture Learning (CAOCL); Colonel Jeffery W. Bearor, USMC (Ret), SES TECOM and previous director of CAOCL; Lieutenant Colonel William D. Shannon, CAOCL; Colonel Steven M. Zotti, Marine Corps Strategy and Vision; and Dr. Kerry B. Fosher, Marine Corps Intelligence Agency.

The first chapter in this book, by Major Jonathan P. Dunne, was initially published in the *Marine Corps Gazette*. We thank the

Gazette for permission to revise and include the chapter here. The staff at the Library of the Marine Corps also provided much needed-assistance in locating source material and obtaining copyrights. We are especially grateful to Ms. Carol E. Ramkey, director of the Library of the Marine Corps, and Ms. Susan E. Warren, copyright clerk, Library of the Marine Corps. We would also like to thank Mr. Raymond Boisvert for his diligence and attention to detail editing the source document information.

Finally, each of the contributors to this volume would like to thank our spouses and families for the immense time, patience, and support they have offered during the writing and editing of the chapters for this volume. Without their support and encouragement, none of this would have been possible.

Introduction

From Capitol Hill to the military recruit depots, "culture" has become a key issue in planning for U.S. military operations in the 21st Century. Today, congressional testimony is likely to include discussions of the people of Iraq, as well as information on the latest weapons system.[1] Similarly, military pre-deployment training is likely to include classes on Afghan culture and Islam, as well as traditional range or artillery training.

Indeed, culture training is no longer exclusively offered to military units deploying to Iraq and Afghanistan. Military culture programs now extend from recruit training and Officers' Candidate School to Professional Military Education (PME) and cover a wide spectrum of functions: from Civil Affairs to Special Operations; from Military Training Teams to Reconstruction and Stabilization Operations; from Information Operations to Intelligence; and from Logistics to Ground Operations. In accordance with the goals of the Marine Corps Vision 2025,[2] within the next 15 years, the Marine Corps will provide culture, regional, and language training and education to all Marines throughout the fleet. Currently, culture has been included in the latest *Marine Corps Training and Readiness Manual*[3] and is being written into Marine Corps doctrine, as this book goes to press.

Reflecting the growing emphasis on understanding cultural factors, the Marine Corps, the Army, the Air Force, and the Navy have each created culture centers mandated to provide culture training and educational materials to their respective services.[4] In particular, the Marine Corps University (MCU) and the Center for Advanced Operational Culture Learning (CAOCL) have developed an operational culture curriculum spanning PME from Lieutenant to Lieutenant Colonel. The Air Force's Air University is developing and implementing a similarly ambitious plan for a professional military culture education program at its institution. This rapid increase in cultural programs is an important step forward in the Marine Corps' and the U.S. military's efforts to effectively operate in irregular and conventional warfare environments.

Nearly 70 years ago, the Marine Corps' *Small Wars Manual* noted the importance of understanding and studying the local people in order to succeed in small wars:

> The motive in small wars is not material destruction. It is usually a project dealing with the social, economic, and political development of the people. It is of primary importance that the fullest benefit be derived from the psychological aspect of the situation. This implies a serious study of the people, their racial, political, religious and mental development.[5]

Despite this advice, however, only recently have the Marine Corps and the U.S. military in general paid serious attention to the development of a serious culture education program. This has not meant that Marines and other service members have neglected cultural factors in their operations in the past. Rather, Marines have simply developed ad hoc solutions to complex cultural challenges in the field. While some of this "learning by doing" has filtered down over the decades, sadly, much of this knowledge has been lost. As a result, one of the greatest challenges for the rapidly developing contemporary culture programs is capturing, analyzing, and publishing materials on operational culture that can be used in the military classroom.

Today, due to repeated deployments in Iraq, Afghanistan, and around the globe, the Marine Corps has one of the most culturally astute generations of fighters. Over the past few years, the Marine Corps University has been fortunate to include many of these highly experienced officers in its year-long professional military education programs. Recognizing the importance of capturing the knowledge and experience of these officers, the editors of this book have assembled a set of six master's papers on culture and military operations written by officers while they were in residence at MCU. The collection of readings that follows provides on-the-ground field experience and perspectives of five U.S. Marine officers and one Australian Army officer while at Command and Staff College and the School of Advanced Warfighting from 2007 to 2009.

These officers have worked with local populations in foreign countries around the globe (including Iraq, Afghanistan, and Pakistan). Significantly, these authors not only possess an in-depth understanding of the cultural issues in their areas of operations, but also have the rigorous theoretical background to make sense of their experiences on a more global scale. This combination of academic rigor and military expertise provides the audience with a unique set of readings that apply not only to the current operating situation, but also serve as a guide for understanding and preparing for future conflicts.

If history provides any indication of future operations, Marines, along with other U.S. and allied military services, will continue to work with and among foreign populations—an observation supported by the recent publication of the *Marine Corps Strategy and Vision 2025*.[6] By capturing the wisdom and experiences of our current military leaders, perhaps we can avoid the military historical tradition of learning (and re-learning) cultural lessons by trial and error—and instead provide concrete lessons from previous experience to guide future actions.

Operational Culture and the Marine Corps: A Historical Perspective

It is doubtful that there is a military service that has not been required—at some point in its history—to interact with foreign people and understand their culture. Due to its expeditionary nature, the Marine Corps has certainly had some of the longest and most in-depth experience in interacting with foreign populations of any U.S. service.[7] For more than 200 years, Marines have undertaken countless small-scale missions in remote places around the globe, where understanding and working with the local people has been a key to success.

From the Barbary Coast to the Banana Wars, from the recent tsunami in Indonesia to the al-Anbar Awakening in Iraq, Marines have had to learn foreign languages, operate among and alongside local people, and train and deploy with foreign militaries. During the Banana Wars in the early 1900s, for example, many Marines became fluent in Spanish in order to understand the local people in Guatemala, Haiti, and Nicaragua. As a result of its experience during the Banana Wars, the Marine Corps produced the *Small Wars Manual*.[8] This guide continues to be one of the most important U.S. military references on small wars today.

Perhaps one of the most successful of all operational culture efforts was the Marine Corps CAP (combined action platoons) program during the Vietnam War. In order to work more effectively with the local population, platoons of 10 to 12 Marines were embedded in villages in Vietnam. The platoons provided protection for the villagers. More importantly, however, the CAPs gained the trust and support of the local people through their assistance to the locals by providing medical care, education, and economic development projects. As a result, villages with CAPs were more likely to resist Viet Cong efforts at infiltration and to report Viet Cong activities to Marine units.[9]

For the most part, and by necessity, Marines have had to figure out how to work with local populations on their own. This has often been a long and difficult road, based predominantly on trial and error learning. With the exception of the *Small Wars Manual*, Marines have generally been given very little guidance on how to deal with foreign cultures.

With practice, due to their natural ingenuity, Marines have learned to work with local populations. There is no doubt that Marines returning from Iraq and Afghanistan after two, three, or even more tours are significantly more culturally skilled and competent than they were at the beginning of Operation Iraq Freedom I (OIF I). Through repeated interactions with the local population, many Marines have developed essential skills of negotiation—establishing positive relationships with community leaders, and recognizing and working with competing ethnic and tribal groups in their Areas of Operation.

Many of these skills, however, have come at a high cost: lessons learned from failed negotiations, creating the wrong alliances with ineffective sheikhs, or misunderstandings of tribal and ethnic structures have, at best, resulted in slow and frustrating outcomes. At worst, these lessons have been learned with the more serious price tag of the lives of Marines, other service members, or civilians.

Current U.S. military leaders recognize the value of a more systematic approach to culture and stress the importance of developing operationally effective skills in Marines and other service members before they are deployed. They realize that planning ahead of time for cultural factors in operations is far more efficient than learning from one's mistakes in an after action report.

In order to create successful cultural programs, the U.S. military has faced three major challenges. First, creating cultural analyses that are academically sound yet relevant to military personnel has been a key challenge in the development of educational materials. Equally important has been learning cultural lessons from the experiences of those personnel who have returned from the field. Finally, military leaders need guidance on how to effectively prepare our militaries for the cultural challenges of operations in the coming years. The readings in this book seek to address each of these needs.

Creating Sound Cultural Analyses for the Military

Articles on military cultural issues regularly appear in military journals, such as the *Marine Corps Gazette*,[10] the *Military Review*,[11] and *RUSI*,[12] as well as in national magazines, such as *Harper's Magazine*[13] and the *National Journal*.[14] In particular, the Army has produced a culture guide focusing on Iraq,[15] while the Marine Corps has published recently two handbooks on culture and military operations.[16] Using a more psychological approach to culture, the Air Force has begun publishing papers on the concept of Cross-Cultural Competency (3Cs), in a collaborative effort with the Army Research Institute.[17] Additionally, the military intelligence community has produced an entire journal issue devoted to culture.[18]

Although this flurry of activity is encouraging, not all of the materials and programs that have been produced thus far have been of equal quality. In the hurry to push "culture" out to the operating forces, quick, easy-to-digest models of culture are often selected for training. Such models appeal to the military because of their visual simplicity on a power point presentation or their quick transformation into a "check-in-the-box" form. However, when tested in the operating environment—the only test that matters to a military serviceman whose life is on the line—these models do not always deliver.

Models, such as Hofstede's[19] Individual versus Collective Dichotomies or Maslow's Hierarchy of Needs,[20] have been distributed across the services without any consideration for their origins or limitations. Hofstede, for example, is a psychologist who examined cross cultural negotiating styles in international business communities: a far cry from bargaining with a warlord in war-torn Afghanistan.[21] On the other hand, Maslow developed his hierarchy of needs from studies of extraordinary U.S. achievers and college students, not Muslim radicals in Iraq.[22] These scholars' findings were never intended to extend to the military situations in which U.S. Marines and other service members find themselves today.

6   *Applications in Operational Culture*

Squeezing academically solid cultural models into the "wrong bottles" offers one set of problems and flaws. At the other end of the spectrum, however, are tales from the field that are often very culturally specific and are not necessarily applicable to other cultures or regions. While such "lessons learned" provide good case studies, the military needs to avoid the tendency to assume that the cultural lessons of military experiences in one specific Area of Operations can be generalized to other cultures or situations. Recent articles have cautioned against assuming that all tribes behave similarly, that one counterinsurgency is like another, or that there is a "cookie cutter" approach to working with foreign cultures.[23]

Marines and our service partners need cultural approaches that are firmly grounded in both military experience and social science research and concepts. The first two readings in this book provide exactly this approach. Both Major Jonathan P. Dunne and Lieutenant Colonel Alejandro P. Briceno are Marines who have extensive experience operating in foreign military environments and a solid understanding of key academic theories of culture. Major Dunne has deployed twice to Iraq, including a one-year assignment in a Military Transition Team (MTT) as a combat advisor embedded with 2d Brigade, 7th Army Division. As this book goes to print, Major Dunne is headed to AFRICOM, located in Stuttgart, Germany. Lieutenant Colonel Briceno has deployed to Okinawa, South Korea, Iraq, and, most recently, to Stuttgart to help stand up AFRICOM.

Major Dunne's chapter offers a compelling challenge to the assumption that Maslow's Hierarchy of Needs is appropriate for all cross-cultural situations. In his chapter, "Maslow is Non-deployable: Modifying Maslow's Hierarchy for Contemporary Counterinsurgency," Dunne takes a look at Iraqi culture from the perspective of an officer who has spent a long period of time working side by side with the Iraqi military. He argues that "Maslow's hierarchy is founded upon American or Western ideals, not the principles and standards that define Iraq's non-Western culture." He points out that Maslow's model assumes a culture in which individual needs and achievement take priority over the group—a cultural assumption that does not necessarily apply in Iraq. Dunne's critique forces us to re-evaluate the military's tendency to grab the latest, easy check-in-the-box list from an introductory psychology book and apply it to the complex cultural environments in which the military must act.

While Major Dunne cautions the military in transferring its Western models to other cultures, Lieutenant Colonel Briceno seeks to develop new approaches that are theoretically sound and relevant to the military. Using his expertise in computer analysis and programming, Lieutenant Colonel Briceno offers an intriguing new graphic model to help reduce the confusion and quantity of cultural information through which a commander must wade. In his chapter, "The Use of Cultural Studies in Military Operations: A Model for Assessing Values-Based Differences," Briceno proposes a computer analysis that could compare cultural differences on key value categories. This model could assist leaders in quickly identifying those cultural issues (e.g., gender, notions of justice or ethnicity) that are most likely to become friction points between U.S. forces and a specific indigenous population. Commanders could then focus their efforts on training their troops on critical cultural issues, rather than training for every possible cultural factor.

Lessons Learned

Developing and evaluating the models that the military uses to understand cultural factors is one challenge. Equally important, however, is the task of learning cultural lessons from current operations. The U.S. military has many sources for learning from its experiences in foreign areas. In particular, centers for lessons learned are long-standing institutions in the U.S. military. These centers provide extensive debriefs of Marines, soldiers, and other service members as they return from the field.

Military journals also provide an excellent forum to present and discuss the success and failure of current military operations. Yet, despite the rapid proliferation of assessments of current operations, the military's understanding of the role of culture in the successes and failures in Iraq and Afghanistan, or elsewhere, is still vague to say the least. There are several explanations for this surprising gap. First, most lessons learned debriefs are based on traditional questions that focus on conventional operational issues. The interviews tend to emphasize understanding failures in equipment, tactical or procedural issues, and maneuvers. Questions about cultural challenges are haphazard, or nonexistent. Thus, despite the mounds of data on current U.S. military operations, cultural information on current operations is limited and extremely difficult to locate.

There are encouraging exceptions to this pattern, however. Recognizing the need for culture-specific vignettes, a number of organizations have been collecting cases of culture related issues for use in teaching and training. For example, the Joint Center for International Security Force Assistance (JCISFA) has been compiling a large data base of culture stories from personnel returning from Advisor Training Groups. Similarly, the Marine Corps Center for Advanced Operational Culture Learning (CAOCL) has debriefed Marines, and especially foreign area officers (FAO), on culturally focused issues in their deployments; and, the Marine Corps Intelligence Activity has begun a cultural vignettes project, aimed at collecting concrete examples of cultural issues that affect military operations in the field. Currently, however, none of these projects have been published or compiled in a manner that is easily accessible to the general military community.

A second problem in learning cultural lessons from current military experiences is the lack of detailed in-depth case studies. Despite the increasing number of culture related articles in military journals, many of these articles are focused on general principles that can be applied to larger operational issues, such as COIN (counterinsurgency) or SSTR (Security, Stability, Transition, and Reconstruction). While a number of articles provide specific cases or field experiences that include cultural issues, the analysis is more tantalizing than concrete, since understanding the cultural factors at play is not the primary focus of the article.[24]

Yet, detailed case studies are essential to effective cultural learning and analysis. Anthropologists have long recognized that successful understanding of foreign cultures must be based on "deep, thick, rich ethnographies"[25]—or in military terms, thorough, detailed case studies.

A common saying in the Marine Corps is, "The devil is in the details." Nothing could be truer for understanding and learning from operations in foreign cultures. Because of our Western tendency to understand what we see from our own perspective, it is very easy for an outsider to observe an event in a foreign culture and draw completely incorrect conclusions about its meaning and purpose. This is where the details come in. If facts and observations are carefully collected, in an after action report or analysis, details that were initially neglected as insignificant can take on great importance. Without these details, the analysis simply becomes another affirmation that "those Iraqis/Afghanis/Sudanese/etc., don't make any sense."

Major John E. Bilas' chapter, "Developing the Iraqi Army: The Long Fight in the Long War," offers the kind of detailed case study necessary for accurate analysis of cultural factors in military operations. Major Bilas is a Marine Corps intelligence officer who served as a military transition team (MTT) advisor for the Iraqi Army from 2006-2007 with 2d Brigade, 7th Army Division. He currently oversees pre-deployment training, including the cultural exercises, at Mojave Viper in Twentynine Palms, California. Major Bilas' chapter provides a careful case study of the social and cultural challenges of working on an Iraqi military transition team. What makes his analysis so significant is the immense detail he provides regarding the many logistical problems his team and successive MTTs faced in working with the Iraqi Army. He uses a thorough methodology of comparing information from interviews, after action reports, official documents, and personal observations to create an extremely rich and in-depth evaluation of the issues. Through such careful, painstaking research, the reader is able to develop a clear appreciation for the social and cultural challenges the MTT experienced. Major Bilas' chapter offers many practical lessons to both future MTT leaders and other military personnel planning to work closely with non-Western militaries.

As the U.S. military shifts its energies toward Afghanistan, Major Randall S. Hoffman's chapter, "The Way Ahead: Reclaiming the Pashtun Tribes through Joint Tribal Engagement," offers a timely and critical cultural perspective on operations in the region. Major Hoffman's intensive experience working alongside Pashtuns in Afghanistan as a military advisor from 2003 to 2005 serves as the basis of his argument that the varied Pashtun tribal groups in Afghanistan are the center of gravity in the conflict between the Afghan state and al-Qaeda/Taliban forces. Combining careful historical research, ethnographic studies of the Pashtun, personal interviews with Pashtun tribal leaders, and his own field experiences, Hoffman provides a compelling argument for focusing contemporary U.S. military operations in Afghanistan on gaining the support of Pashtun tribal groups—particularly the Ghilzai and Karlanri Pashtun that inhabit the eastern areas of Afghanistan on the border of Pakistan.

Major Hoffman does not restrict his analysis to Afghanistan, but focuses on the important role of the Pashtun living across the border in Pakistan's North-West Frontier Province (NWFP) and the Federally Administered Tribal Areas (FATA). His cross-state conceptualization of the problem emphasizes the difference between a politically oriented approach that defines conflict in terms of a centralized state with defined boundaries and a cultural/ethnic/tribal analysis that evaluates conflict from the identities and perspectives of the actors, regardless of their location on a geographic map.

Planning for the 21st Century

Lieutenant Colonel Briceno's and Major Dunne's chapters develop and evaluate the cultural models and theories used by the military in order to better conceptualize the issues. Majors Bilas and Hoffman use detailed case studies to evaluate the specific cultural contexts of contemporary conflicts, providing concrete applications of theory to field situations in Iraq and Afghanistan. The final two readings, by Major Castro and Lieutenant Colonel Brain, look toward future military operations, offering recommendations for ways to prepare for the cultural challenges ahead.

Major Robert T. Castro, like Major Hoffman, views local populations as a key center of gravity in irregular warfare and counterinsurgency environments where both sides in a conflict vie for the support and assistance of the people. Drawing upon the extensive work of COIN theorists, Castro argues that cultural understanding is not simply a "nice to have" skill to be added to an already full training pack, but a core competency that is essential for success in unconventional warfare environments, where the ability to win the support of the local population may determine the outcome of the conflict.

His argument derives in part from his own challenges working with the local population in Fallujah, Iraq. Major Castro deployed twice with II MEF Headquarters Group to Operation Iraqi Freedom. During his second deployment, he held the operational billet of Mayor of Fallujah, overseeing reconstruction assistance to build and repair the city's infrastructure. As a result of his experiences, Castro proposes that leadership, anthropological, and cultural skills cannot simply be developed as required in the field, but must be formally built throughout the training and education continuum. His chapter proposes a career-long, cultural education and training program from boot camp and Officers' Candidate School to advanced general officers' courses.

Major Castro's proposed curriculum progression goes far beyond theory, however. Currently, the Marine Corps Center for Advanced Operational Culture Learning in Quantico, Virginia, is working with all branches of the Marine Corps to develop and implement such a wide-ranging and extensive culture curriculum within the next decade.

The U.S. Marine Corps is not the only military service to struggle with the challenges of incorporating culture into education and training. This issue is at the heart of Lieutenant Colonel Steven Brain's chapter, "Operational Culture: Is the Australian Army Driving the Train or Left at the Station?" Lieutenant Colonel Brain's chapter draws upon over 16 years' experience as an officer in the Australian Army, during which he has deployed to Rwanda, East Timor, and Iraq. Noting that the Australian Army has tended to leave cultural training up to the commander in an ad hoc manner, Lieutenant Colonel Brain argues for a serious reevaluation of the way the Australian Army conducts culture training. In order to successfully incorporate culture into operations, he proposes that the Australian Army must change its own military structure and training methods.

By shifting the focus from cultures "out there" to one's own military culture, Lieutenant Colonel Brain provides an important lesson for all militaries. That is, most of the current focus on culture education and training in the U.S. military has emphasized the need to provide more accurate education and information on foreign cultures. Yet, ultimately, the success of this new culture venture lies in our own military's ability not only to shift its thinking about the nature of warfare, but also to adjust its own structures, processes, and modus operandi in teaching, training, and operating in foreign locations.

Such a shift demands a serious effort to capture and learn from the deep experience and perspectives of our current operating fleet. The readings in the following chapters are each a step forward in this direction. If this generation of culturally-skilled military personnel can pass on its wisdom, perhaps one day, the Marine Corps and our sister services will no longer need to rely on trial and error to operate successfully in other cultures—and that certainly would be a meaningful change in the way that our own military culture does business.

Notes

[1] Gen David H. Petraeus (USA), "Report to Congress on the Situation in Iraq," 8-9 April 2008 (http://www.mnf-iraq.com/images/stories/Press_briefings/2008/april/080408_petraeus_testimony. pdf, accessed 10/15/08).

[2] Commandant of the Marine Corps, *Marine Corps Vision and Strategy 2025* (Washington, DC: U.S. Government Printing Office, 2009).

[3] USMC Ground Training Branch, *Marine Corps Training and Readiness Manual*, NAVMC 3500.65 *Operational Culture and Language* (https://www.intranet.tecom.usmc.mil/sites/gtb/products/manuals/default.aspx; access limited to Department of Defense CAC card holders).

[4] The Marine Corps Center for Advanced Operational Culture Learning (CAOCL) in Quantico, VA; the Army TRADOC Culture Center (TCC) in Fort Huachuca, AZ; the Air Force Culture and Language Center (AFCLC) in Montgomery, AL; and the Navy Center for Language, Regional Expertise and Culture (CLEREC) in Pensacola, FL.

[5] U.S. Marine Corps, *Small Wars Manual*, FMFRP 12-15 (Washington, DC: U.S. Government Printing Office, 1940), 1-10.

[6] *Marine Corps Vision and Strategy 2025*.

[7] Sgt Kurt M. Sutton (USMC), "A Look at the Marine Corps' Past," *Marines Magazine*, February 1997.

[8] USMC, *Small Wars Manual*.

[9] Michael E. Peterson, *The Combined Action Platoons: The U.S. Marines' Other War in Vietnam* (New York: Praeger, 1989).

[10] See, for example, Maj Jonathan Dunne (USMC), "Twenty-Seven Articles of Lawrence of Arabia," *Marine Corps Gazette,* October 2007, 67-70; Paula Holmes-Eber and Barak A. Salmoni, "Operational Culture for Marines," *Marine Corps Gazette*, May 2008, 72-77; Lieutenant Colonel James B. Higgins (USMC), "Culture Shock: Overhauling the Mentality of the Military," *Marine Corps Gazette*, February 2006, 48-51; SSgt Rory O'Hara (USMC), "ASCOPE: Planning Acronym for a New Generation of Warfare," *Marine Corps Gazette*, January 2007, 46-47; Capt Michael C. Vasquez (USMC), "Tribalism Under Fire: A Reexamination of Tribal Mobilization Patterns in a Counterinsurgency," *Marine Corps Gazette*, January 2008, 62-67.

[11] See, for example, Henri Bore, "Cultural Awareness and Irregular Warfare: French Army Experience in Africa," *Military Review*, July-August 2006, 108-11; Lieutenant Colonel Michael Eisenstadt (USAR), "Tribal Engagement: Lessons Learned," *Military Review*, September-October 2007, 16-30; Maj Remi Hajjar, (USA), "The Army's New TRADOC Culture Center," *Military Review*, November-December 2006; Pauline Kusiak, "Sociocultural Expertise and the Military: Beyond the Controversy," *Military Review*, Nov-Dec. 2008, 65-76; Maxie McFarland, "Military Cultural Education," *Military Review*, March-April 2005, 62-63; Barak A. Salmoni, "Advances in Pre-Deployment Culture Training: The U.S. Marine Corps

Approach," *Military Review*, November-December 2006, 79-88.

[12] LtGen Sir John Kiszely, "Learning About Counter-Insurgency," *RUSI Journal* 152 (December 2006).

[13] Steve Featherstone, "Human Quicksand: For the U.S. Army, a Crash Course in Cultural Studies," *Harper's Magazine*, September 2008, 60-68.

[14] Sydney J. Freedberg, "Chess with the Sheiks," *National Journal*, 12 April 2008, 26-31.

[15] William D. Wunderle, *Through the Lens of Cultural Awareness: A Primer for US Armed Forces Deploying to Arab and Middle Eastern Countries* (Fort Leavenworth, KS: Combat Studies Institute Press, 2006).

[16] Barak A. Salmoni and Paula Holmes-Eber, *Operational Culture for the Warfighter: Principles and Applications* (Quantico, VA: Marine Corps University Press, 2008); USMC Marine Corps Intelligence Activity, C-GIRH: Cultural Generic Information Requirements Handbook, MCIA-7634-001-08, August 2008.

[17] Allison Abbe, Lisa M.V. Gulick, and Jeffrey L. Herman, *Cross Cultural Competence in Army Leaders: A Conceptual and Empirical Foundation* (Arlington, VA: U.S. Army Research Institute for the Behavioral and Social Sciences, Study Report 2008-01, October 2007; online at http://www.hqda.army.mil/ari/pdf//SR%5F2008-01.pdf); Brian R. Selmeski, Brian R, *Military Cross-Cultural Competence: Core Concepts and Individual Development* (Royal Military College of Canada, Centre for Armed Forces & Society, Occasional Paper Series #1, 16 May 2007).

[18] See, for example, Col Bryan N. Karabaich (USA, ret), and Jonathan D. Pfautz, "Using Cultural Belief Sets in Intelligence Preparation of the Battlefield," *Military Intelligence Professional Bulletin PB 34-06-02* (April-June 2006), 40-49; Scott Swanson, "Asymmetrical Factors in Culture for SOF Conflicts: Gaining Understanding and Insights," *Military Intelligence Professional Bulletin PB 34-06-02* (April-June 2006), 29-33; LtCmdr Sylvain Therriault (Canadian Forces) and Master Warrant Officer Ron Wulf (Canadian Forces), "Cultural Awareness or If the Shoe Does not Fit . . ." *Military Intelligence Professional Bulletin PB 34-06-02* (April-June 2006), 22-26.

[19] Geert Hofstede, *Cultures and Organizations: Software of the Mind* (Cambridge, UK: McGraw-Hill, 1991).

[20] Abraham H. Maslow, *Motivation and Personality* (1954; reprint, New York: Harper-Collins, 1987).

[21] See Salmoni and Holmes-Eber, *Operational Culture for the Warfighter*, for a critique of Hoftstede's applicability to military situations.

[22] For critiques of Maslow's relevancy in other parts of the world, see Manfred A. Neef, *Human Scale Development Conception Application and Further Reflections* (New York: Apex Press, 1989).

[23] LtCol Michael Eisenstadt (USAR), "Tribal Engagement: Lessons Learned," *Military Review*, September-October 2007, 16-30; John Nagl, *Learning to Eat Soup with a Knife: Counterinsurgency Lessons from Malaya to Vietnam* (Chicago: University of Chicago Press, 2004); Maj Christopher H. Varhola (USAR) and Lieutenant Colonel Laura R. Varhola (USA), "Avoiding the Cookie-Cutter Approach to Culture: Lessons Learned from Operations in East Africa," *Military Review*, December 2006, 73-79; Capt Michael C. Vasquez (USMC), "Tribalism under Fire: A Reexamination of Tribal Mobilization Patterns in a Counterinsurgency," *Marine Corps Gazette*,

January 2008, 62-67.

[24] See, for example, Maj Niel Smith (USA) and Col Sean MacFarland (USA), "Anbar Awakens: The Tipping Point," *Military Review*, March-April 2008, 41-52; Lieutenant Colonel Gregory Wilson (USA), "Anatomy of a Successful COIN Operation: OEF Philippines and the Indirect Approach," *Military Review*, November-December 2006, 2-12; LtCol Leonard DeFrancisci (USMCR), "Money as a Force Multiplier in COIN," *Military Review*, May-June 2008, 21-29.

[25] Clifford Geertz, *Interpretation of Cultures* (New York: Basic Books Classics, 2000).

[26] For a promising new direction in developing concrete military cultural cases, see the work coming out of the Army's Human Terrain Systems, for example, Dave Matsuda, "Human Terrain Teams, Military Ethnography, and how Operational Cultural Knowledge is Changing the Nature of Warfare" (paper presented at the 2008 IUS Conference, Kingston, Ontario, Canada, 7-9 November 2008).

Chapter 1

Maslow is Non-Deployable: Modifying Maslow's Hierarchy for Contemporary Counterinsurgency[1]

Major Jonathan P. Dunne, USMC

Major Jonathan P. Dunne is a graduate of the University of Rochester and was commissioned as a U.S. Marine in 1993. As a second lieutenant, he became an artillery officer and then served as a platoon commander and fire direction officer with Battery B, 1st Battalion, 10th Marines. As a first lieutenant, he served as a team officer-in-charge of a 2d ANGLICO detachment. As a captain, he served as the inspector-instructor of Battery P, 5th Battalion, 14th Marines, in Spokane, Washington. At his current rank, Major Dunne served as the fire support coordinator, operations officer, and then executive officer of 3d Battalion, 11th Marines.

Major Dunne has deployed twice in support of Operation Iraqi Freedom, including a one-year assignment as a combat advisor embedded with 2d Brigade, 7th Iraqi Army Division. It was Major Dunne's experiences during these two deployments, as well as the operational culture courses of instruction at the Marine Corps Command and Staff College (AY 07-08) and the School of Advanced Warfighting (AY 08-09), that serve as the foundation of his essay. Major Dunne is currently assigned to the G-5 at Marine Corps Forces, Africa.

As Operation Iraqi Freedom (OIF) evolved, so, too, did the Marine Corps' cultural-training efforts. Initially, cultural sensitivity training (i.e., cursory etiquette instruction) was employed in 2003, to prepare Marines for the Invasion of Baghdad during OIF I. By 2004, when the 1st Marine Division returned to Iraq for OIF II, training had evolved into culture awareness classes (i.e., introductory history, politics, religion overview). It was not until 2005 that training progressed into today's operational culture learning.[2]

Statement of the Problem

As the Marine Corps' professional military education (PME) requirements move away from cultural awareness and begin to take on an operational culture approach, many models were developed to help Marines apply cultural theories in the field. Many of these models, however, are based on elementary sociological or psychological theories—theories that were developed by Western social scientists to explain Western behaviors. One theory in particular, Maslow's Hierarchy of Needs[3], which was developed in the 1940s, has been used Marine Corps-wide to help Marines understand the operating environment in Iraq. However, this model is individual-centric as opposed to group-centric (i.e., a humanistic psychology model) and looks at other cultures from a Western perspective. Using such a prescriptive model that does not take into account Iraqi motivations and ideals may not yield the result the Marine Corps is seeking. The question then becomes: How can the Marine Corps effectively educate its personnel about Iraqi culture without producing overly reductive theories and models?

Operational Culture Defined

Operational culture learning has become central to the way the Marine Corps conducts counterinsurgency operations, which has led to development of many operational culture programs within the Corps' professional military education programs, including the establishment of a Center for Advanced Operational Culture Learning (CAOCL) at the Marine Corps University. Dr. Barak A. Salmoni, deputy director of the center from 2005 to 2008, and Dr. Paula Holmes-Eber, professor of operational culture at the university, collaborated on the publication of *Operational Culture for the Warfighter: Principles and Applications*. This educational publication, based on both historical examples and contemporary case studies, provides a useful framework from which to view a tactical counterinsurgency (COIN) problem through a cultural lens.

In their book, Salmoni and Holmes-Eber define operational culture as: "those aspects of culture that influence the outcome of military operation; conversely, the military actions that influence the culture of an area of operations."[4] The definition of operational culture is outlined in Figure 1 to the right.[5] This article will examine the application of operational culture and will critique one of the most frequently used operational culture models—Maslow's Hierarchy of Needs—to highlight problems associated with applying Western-perspective models and to emphasize the need for the Marine Corps to review its current counterinsurgency doctrine.

Figure 1

CAOCL's Operational Culture Construct

Operational Culture – Defined:
Those aspects of culture that influence the outcome of military operation; conversely, the military actions that influence the culture of an area of operations.

Dimensions of Operational

PHYSICAL ENVIRONMENT	All cultures have developed a unique interdependent relationship with their **physical environment**.
ECONOMY	All cultures have a specific system for obtaining, producing, and distributing the items that people need or want to survive in their society. This system (which does not necessarily require money or banks) is called the **economy of culture**.
SOCIAL STRUCTURE	All cultures assign people different roles, status, and power within a group. The way people organize themselves and distribute power and status is called their **social structure**.
POLITICAL STRUCTURE	All cultures have a system that determines who leads the group and makes decisions about its welfare. How a group is ruled (and it may not be by a specific person or set of people) is referred to as the **political structure** of a culture.
BELIEF SYSTEM	All cultures have a shared set of **beliefs and symbols** that unite the group.

18 *Applications in Operational Culture*

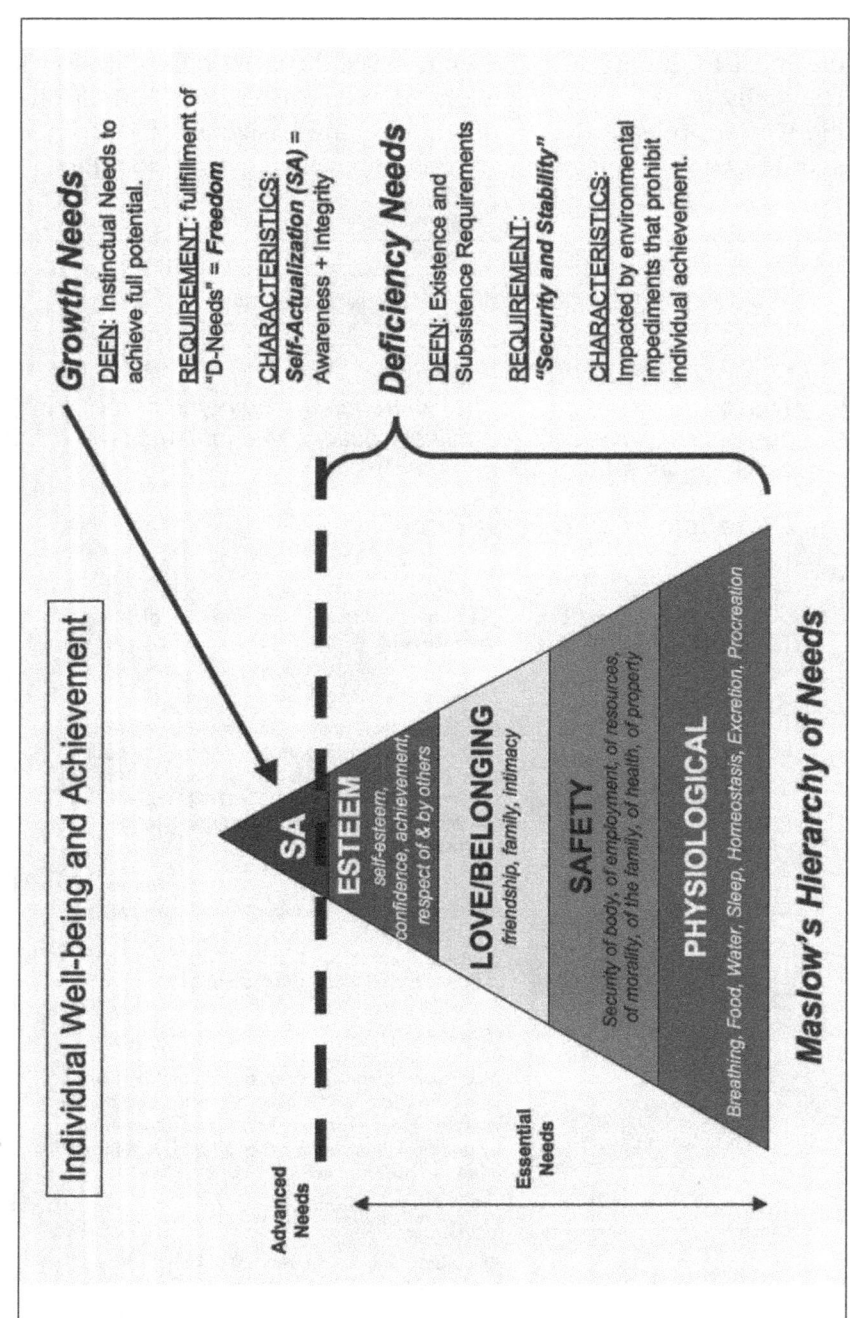

Figure 2

Understanding Ourselves through Maslow

Abraham H. Maslow (1908–1970), an American psychologist, is considered the father of humanistic psychology. He is best remembered for his 1943 paper, *A Theory of Human Motivation*, which outlines his theory regarding the hierarchy of human needs.[6] Commonly referred to as "Maslow's Hierarchy of Needs," Maslow's construct provides a simplistic model for examining the necessities and motives of human beings.

Figure 2 to the left depicts Maslow's Hierarchy. The four lowest layers of the hierarchy's pyramid, referred to as "Deficiency-Needs" or "D-Needs," are those human requirements necessary to sustain life. The most basic needs, which form the foundation of Maslow's pyramid, are *physiological needs*, such as food, water, and sleep (Figure 1, the base of the pyramid). The next need-priorities, in order of precedence, are *safety*, the provision or feeling of *love and belonging*, and *self-esteem*.[7] Maslow argued that these four D-needs were conditions or requirements to sustain human life and could be negatively affected by environmental impediments.[8]

Once D-needs are adequately provided for, individuals will attempt to fulfill more advanced, self-defined *growth needs*, which is an instinctual need to achieve one's full potential.[9] If an individual is able to fulfill both his D-needs and growth needs, he will have achieved *self-actualization*, or the tip of the pyramid—a condition that is contingent upon individual freedom and defined by one's awareness, integrity, and achievement of full-potential. According to Maslow's research in the U.S., Americans, who have been raised in a culture emphasizing individualism over group affiliation, typically prioritize their needs as Maslow defines them within his hierarchical-pyramid.

The Iraq Case: Reassessing Maslow

A problem arises when applying Maslow's construct to comprehend contemporary Iraqi culture. Maslow's Hierarchy is founded upon American or Western ideals, not the principles and standards that define Iraq's non-Western culture. More specifically, Maslow's Hierarchy of Needs does not adequately address the *cultural* differences between United States Marines and Iraqi soldiers and citizens because the theory emphasizes the needs of the individual

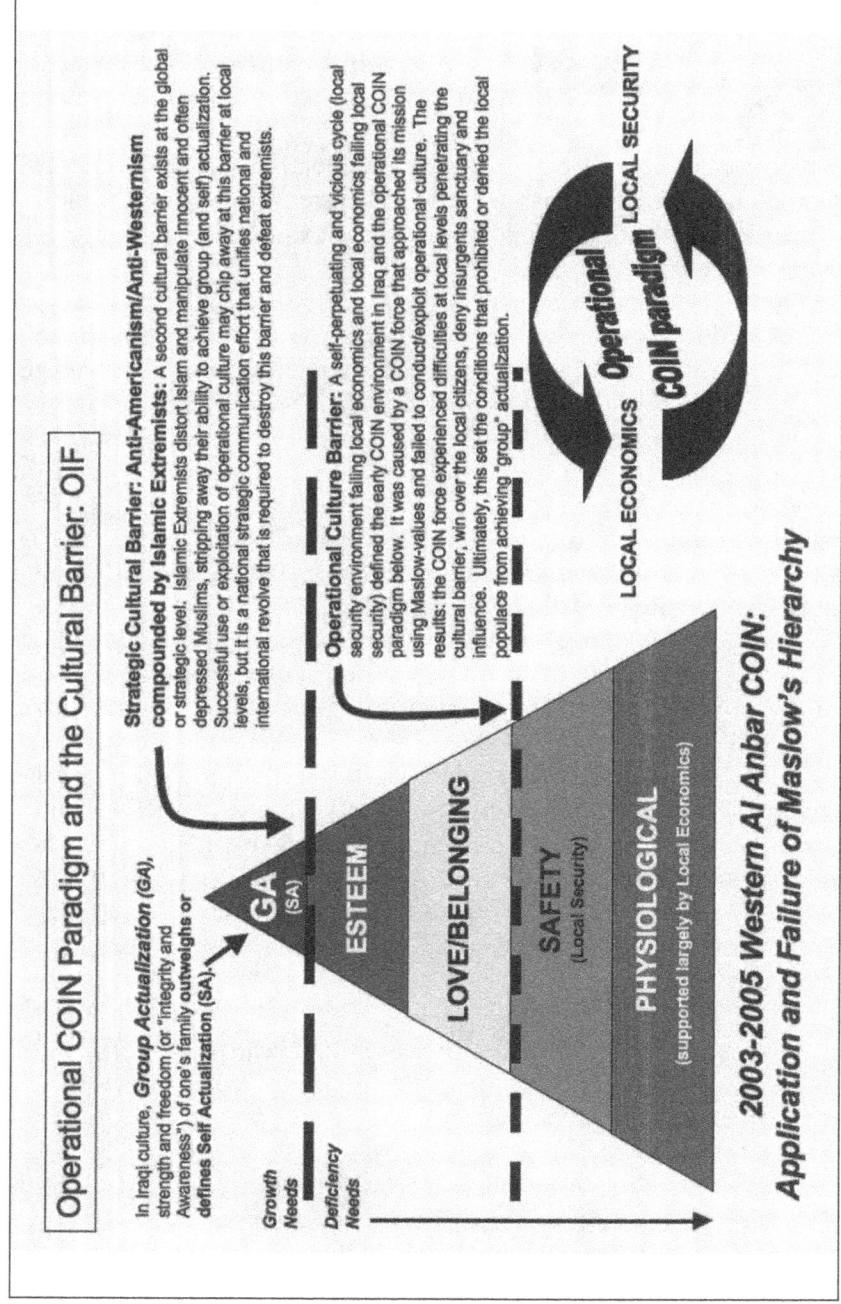

Figure 3

(an American cultural value) before the requirements of the group (i.e., family, clan, tribe in which an individual is a responsible, functioning member). In Iraq, the cultural value of the collective is a deeply entrenched belief. Metaphorically speaking then, Maslow's construct within Iraqi culture[10] puts the cart before the horse–it places self-needs before collective demands. Thus, Maslow's theory is "non-deployable."

The Marine Corps' delay in developing, promoting, and employing operational culture resulted in its inability to effectually penetrate the existing "operational cultural barrier," as defined in Figure 3. As a result, at least within the Western al-anbar Province during OIF II in 2004 and early 2005, coalition efforts approached a culminating point[11] long before a stable environment was created. This culminating point resulted from a self-perpetuating cycle of sub-standard security and feeble economic growth that continued to fuel the insurgency.[12] An inadequate security environment prevented micro-economic growth; meanwhile a stagnant local economy influenced potentially neutral or pro-coalition Iraqi citizens to succumb to the pressure of aiding the insurgency in order to feed their families.[13] This cycle of security-failing-economics and economics-failing-security[14] continued to feed itself and resulted in an operational COIN paradigm—the military's inability to enable Iraqi citizens to fulfill Iraqi D-needs (as defined by Maslow), and, ultimately, to achieve Iraqi *group-actualization* (the strength and freedom [or "integrity and awareness"] of one's family, clan, or tribe), which outweighs or defines an individual's *self-actualization*.[15]

Figure 3 to the left illustrates this operational COIN paradigm, in which the coalition endeavors to fulfill Iraqi physiological needs (micro-economics based) and provide safety (security operations), were not mutually supportive or synergistic, preventing COIN efforts from penetrating the more elevated yellow and green tier D-Needs (i.e., love/belonging and self-esteem) and the growth need of self-actualization.

After initial COIN struggles in 2004 and 2005, the Marine Corps refined its operational plans and tactical actions to better suit the operating environment. One of the measures the Marine Corps undertook in 2005 was to recognize that executing operations with an operational culture mindset was crucial. A modified Maslow-style

22 *Applications in Operational Culture*

pyramid, which is depicted in Figure 4, reflects the five dimensions of operational culture as its foundation. Counterinsurgency operations that embrace the five dimensions of operational culture lend themselves to the development of positive understanding, rela-

Figure 4

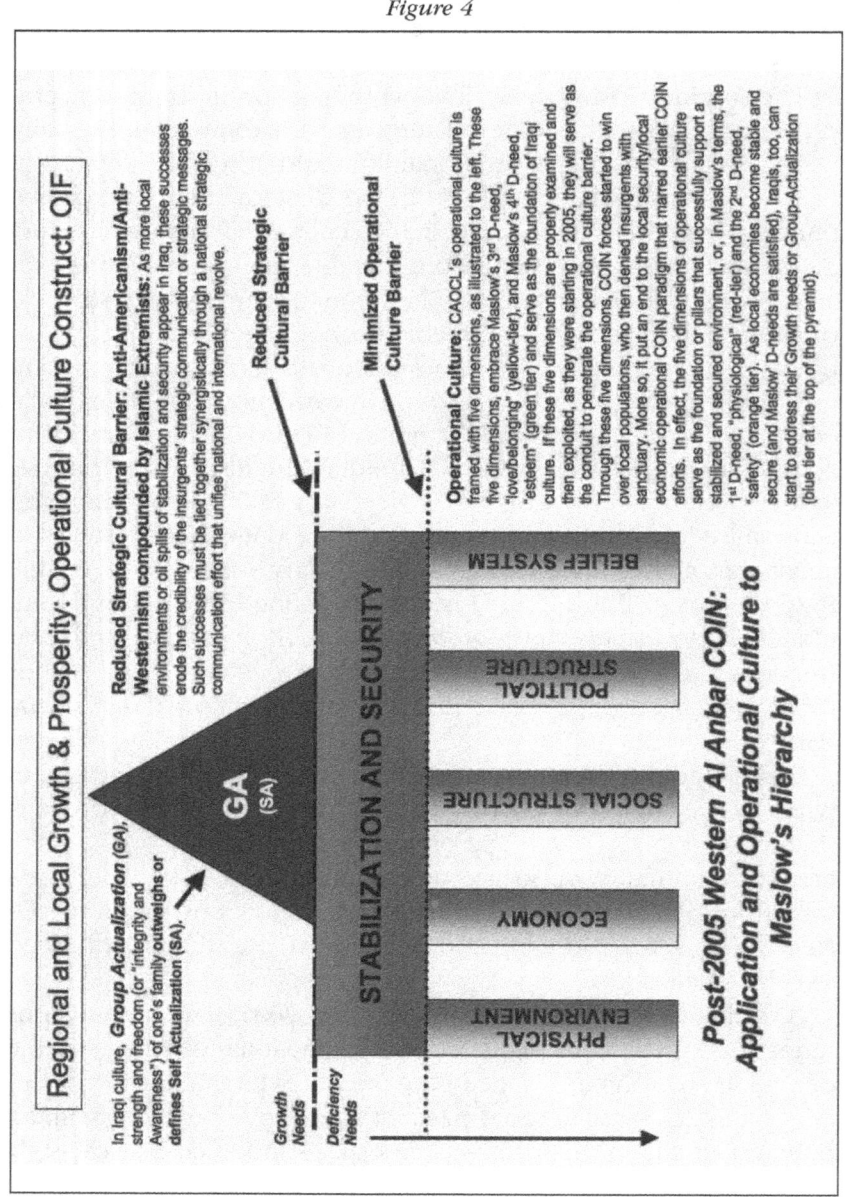

tionships, and mutual trust between Marines and local indigenous populations.

The five pillars, representing the five dimensions of operational culture, promote Maslow's love-belonging and self-esteem D-needs and create an environment that enhances stabilization and security. When an operational culture approach is applied, the problems of the COIN paradigm—as revealed in early US operations in Iraq—can be prevented. "Stabilization and Security" (in Figure 4), supported by the five pillars (i.e., operational culture dimensions), is comprised of the first two Maslow D-needs. The five operational culture pillars embrace Maslow's third tier (love/belonging) and fourth tier (self-esteem). In al-Anbar, once local stability and security were established, families and tribes were able to pursue group actualization, the blue tip-of-the-pyramid in Figure 4. Therefore, the five dimensions of operational culture, which contain the more advanced Maslow D-needs, provide the foundation for achieving the more basic Maslow D-needs (stability and security). Once Marines viewed the Iraq-COIN fight, absent of a traditional American-Maslow lens, they were able to better identify with the local population and win the local's support.

Discussion of the Findings

What It (Operational Culture) All Means

In practice (i.e., during pre-deployment training and also in the conduct of Intelligence Preparation of the Battlefield (IPB) prior to and during deployments), commanders embrace operational culture and integrate it within COIN operations. As a matter of protocol, however, current doctrine marginalizes certain aspects of culture. While Chapter 3 of Marine Corps Warfighting Publication (MCWP) 3-33.5, *Counterinsurgency*, provides useful guidance regarding intelligence efforts within a COIN environment, Chapter 5 of MCWP 2-12, *MAGTF Intelligence Production and Analysis*, is inadequate.

Chapter 5 of the COIN publication outlines the mechanical process of intelligence preparation of the battlespace (IPB). As part of the "Analyze the Battlespace" step of IPB, a miscellaneous, catch-all category, "Other Characteristics of the Battlespace," suggests that vital issues of population demographics, economics, and

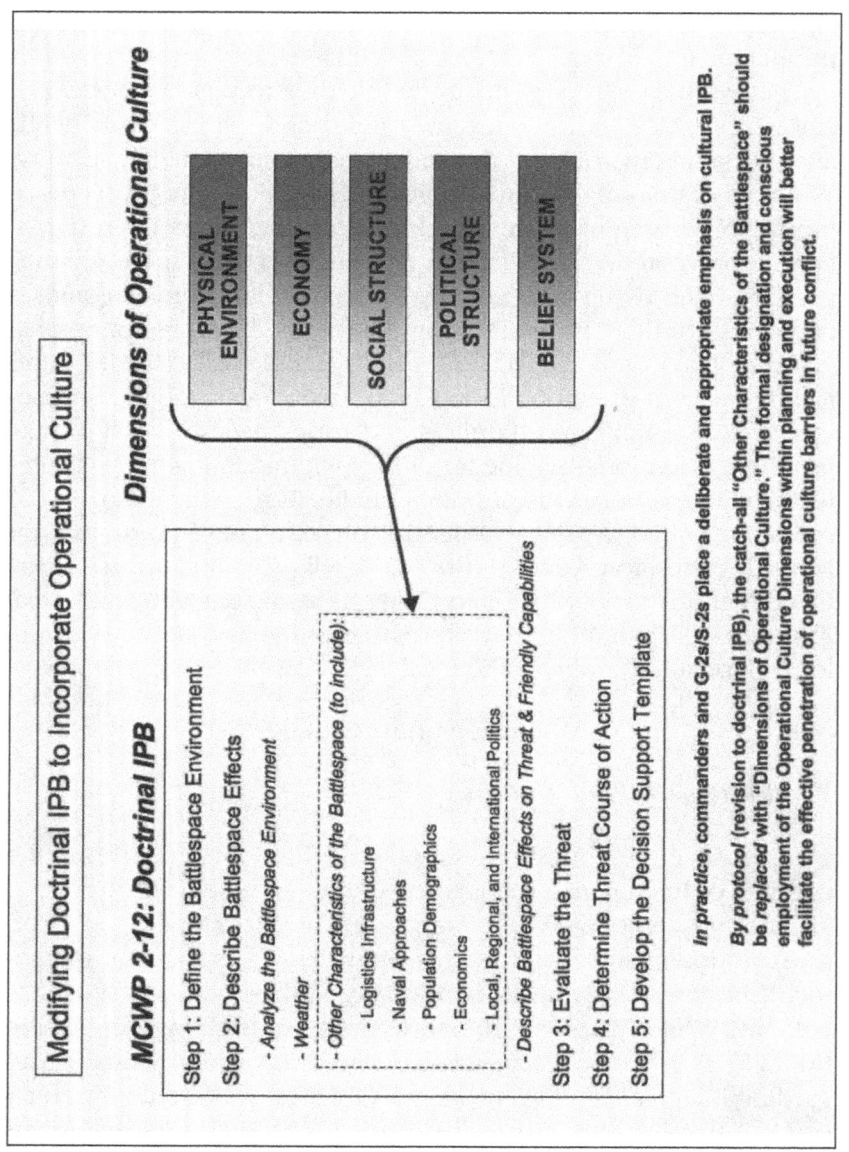

Figure 5

politics should be considered. Contemporary COIN operations in Iraq clearly indicate foreign cultures should not be an afterthought, but should serve as the foundation of intelligence analysis and should permeate all aspects of operational design and planning. Figure 5 illustrates that operational culture and its five dimensions should replace the existing "Other Characteristics" of current, doctrinal IPB. More so, this cultural approach to examining an operating environment should garner greater emphasis within doctrinal planning.

Additionally, Appendix B of Salmoni and Holmes-Eber's *Operational Culture for the Warfighter* contains nearly 150 questions that can be categorized according to the five dimensions of operational culture.[16] Commanders, intelligence officers, and individual Marines should incorporate Appendix B as a foundation in order to successfully formulate information requirements (IRs), define their collections plans, and, understand their operating environment. The formal designation and conscious employment of the Operational Culture Dimensions within planning and execution will better facilitate the effective penetration of operational culture barriers in future conflict.

What It (Operational Culture) Doesn't Mean

The reader may equate the dissolving operational culture barrier in Iraq with the "Great Awakening"[17] that occurred in the al-Anbar Province. To directly link the application of operational culture as the cause for al-Anbar's Great Awakening would be a gross-exaggeration and over-simplification; an exhaustive and expansive effort to combat OIF's terribly complex COIN problem was required. Nonetheless, the successes Marines enjoyed in Western al-Anbar have been related to key cultural shifts as described in Figure 4, (the Maslow modification supported with CAOCL's five dimensions of operational culture).[18] Tools, such as operational culture, have helped Marines better assimilate and operate within the Iraqi culture.

Salmoni and Holmes-Eber's five dimensions of culture are operational in nature and have distinct, small-scale, individual, or small-unit tactical application. The dimensions are not designed to degrade and ultimately dissipate "strategic cultural barriers," as depicted in Figure 3 and again in Figure 4. In fact, the strategic cultural barriers are the deep rooted, anti-American and anti-Western sentiments that have been instigated and/or manipulated by insurgents within Iraq and international actors. While OIF COIN-tacticians, who apply operational culture, labor to mitigate this cultural barrier at local levels, this barrier is a ceiling–a global umbrella-obstruction that blankets local communities. Successful use or application of operational culture, creates "oil-spots"[19] of success that chip away at this strategic barrier.

Conclusion

Ill-prepared Marines previously deployed to Iraq without a proper perspective or effective construct within which to interpret and work within foreign cultures. As products of Western culture, most Marines' hierarchy of needs are similar to those depicted in Maslow's construct. However, this construct, developed more than 60 years ago and based on a Western philosophy, fails to take into account cultural differences between U.S. Marines and Iraqi soldiers and citizens. The five dimensions of operation culture, developed by Dr. Salmoni and Dr. Holmes-Eber, provide a useful tool to analyze foreign cultures. Admittedly, this article is largely based on the author's personal experiences in Iraq, but clearly such an analytical approach to culture could be applied to full spectrum operations, in and amongst foreign cultures around the globe. In practice, commanders, staffs, and Marines, have largely adopted operational culture; by protocol, operational culture should be implemented and reinforced within doctrinal IPB, as it should frame information requirements and collection efforts. If nothing else, Marines must recognize that once the line-of-departure is crossed, mission success hinges on the proper understanding of a (foreign) operating environment's collective values system, not on a Marine's individual perceptions and biases.

Notes

[1] An earlier version of this chapter was published in the *Marine Corps Gazette*, February 2008. This chapter is reprinted with permission of *Marine Corps Gazette*, which retains the copyright.

[2] Barak A. Salmoni, "Advances in Predeployment Culture Training: The U.S. Marine Corps Approach," *Military Review*, November-December 2006, 79.

[3] See Abraham H. Maslow, *Toward a Psychology of Being*, 3d ed. (New York: Wiley, 1999), 81-109.

[4] Barak A. Salmoni and Paula Holmes-Eber, *Operational Culture for the Warfighter: Principles and Applications* (Quantico, VA: Marine Corps University Press), 14.

[5] Ibid., 51.

[6] Maslow, *Toward a Psychology of Being*, 12-29.

[7] Abraham H. Maslow, "A Theory of Human Motivation," *Psychological Review* 50 (July 1943): 370-96 (online at http://psychclassics.yorku.ca/Maslow/motivation.htm).

[8] Maslow, *Toward a Psychology of Being*, 81-109.

[9] "Humanistic Psychology," BookRags.com (http://www.bookrags.com/wiki/Humanistic_psychology), accessed 7 November 2007.

[10] Admittedly, the country of Iraq possesses a diverse number of cultures; for the purposes of this paper, the term "Iraqi culture" will be used as a collective term, with the understanding that tactical practitioners would need to identify and understand the intricacies of a local culture and not simply accept the generalizations of "Iraqi culture."

[11] Carl von Clausewitz, *On War*, ed. and trans. Michael Howard and Peter Paret (Princeton, NJ: Princeton University Press, 1984), 528.

[12] U.S. Government Accounting Office, *Rebuilding Iraq: Resource, Security, Governance, Essential Services, and Oversight Issues* (Washington DC: U.S. Government Accounting Office, June 2004), 5.

[13] U.S. Government Accounting Office, *Stabilizing Iraq: An Assessment of the Security Situation* (Testimony for the Subcommittee on National Security, Emerging Threats, and International Relations, House Committee on Government Reform. Statement for the Record by David M. Walker, Comptroller General of the United States, 11 September 2006), 1.

[14] *Measuring Stability and Security in Iraq* (Report to Congress, in accordance with Conference Report 109-72: Emergency Supplemental Appropriations Act: 13 October 2005), 5-26.

[15] While "self-actualization" and "group-actualization" are presented as separate entities, the tactical practitioner must realize that these concepts are not black-and-white, all-or-nothing propositions. Certainly, individuals in the United States, too, identify the needs and desires of organizations or larger entities, and, certainly, Iraqi citizens do possess individual motives that influence the needs and demands of their clans and tribes. It is the author's argument that Maslow's individual needs pyramid does not properly account for the influence of a dominating, collective factor among Iraqis—the effect of the "group" (for example, clans and tribes).

[16] Salmoni and Holmes-Eber, *Operational Culture for the Warfighter*, 274-88.

[17] The period from 2006-2007 in Operation Iraqi Freedom II in al-Anbar Province in Iraq has been referred to as the "Awakening" because of the significant shift in the local population's perceptions of U.S. intentions in the region. Numerous leading sheikhs shifted their alliances to assist U.S. forces in defeating local insurgencies, leading to a significant drop in the number of improvised explosive devices and greater stability and increased security in the region.

[18] See, for example, Maj Neil Smith (USA) and Col Sean MacFarland (USA), "Anbar Awakens: The Tipping Point," *Military Review*, March-April 2008, 41-52; Sydney Freedberg, "Chess with the Sheikhs," *National Journal*, 12 April 2008, 26-31.

[19] Jean Gottman, "Beageaud, Gallieni, Lyautey: The Development of French Colonial Warfare," in Edward Mead Earle, ed., *Makers of Modern Strategy: Military*

Thought from Machiavelli to Hitler (Princeton, NJ: Princeton University Press, 1943), 254. Lyautey (1854-1934), an influential French colonial theorist, coined the term "oil patch" to describe his colonial strategy in Africa. Metaphorically speaking, an oil spot would represent a small, local "colonial success." As France would achieve a series of local successes, the oil spills or oil spots would run together and eventually blanket an entire area of operations.

Chapter 2

The Use of Cultural Studies in Military Operations: A Model for Assessing Values-Based Differences

Lieutenant Colonel Alejandro P. Briceno, USMC

Lieutenant Colonel Alejandro P. Briceno entered the Marine Corps in 1991 after graduating from The Citadel and served as a platoon commander with Company L, 3d Battalion, 2d Marines, 2d Division. He was deployed to Okinawa and South Korea before becoming a platoon commander with 2d Reconnaissance Battalion, 2d Marine Division, where he worked counter-drug operations. After leaving active duty, Briceno took the S-3A position with 3d Force Reconnaissance Company in Mobile, Alabama. In his civilian position as a senior security consultant for IBM, he was responsible for leading security projects and implementing security solutions for a variety of clients, including the Department of Defense, federal and state agencies, private firms, international clients, and education institutions.

Briceno joined 4th Reconnaissance Battalion in 2001 and subsequently deployed to Iraq with Company C. While deployed, Company C was attached to 2d Force Reconnaissance Company under Task Force Tarawa, where Briceno assumed the executive officer position for 2d Force Reconnaissance Company during combat operations in Iraq. From 2007 to 2008, he attended Command and Staff College at Marine Corps University, where he received a masters in military studies. Currently, Lieutenant Colonel Briceno is deployed to Marine Corps Forces, Africa, where he serves as the current operations officer.

The choice of a point of view is the initial act of any culture.
—José Ortegay Gassett

If one wants to understand a person, one has to know what is important to that person. If one wants to know how to interact with that person, then one must be able to compare his values to that person's values. These ideas apply to personal relationships as well as to diplomatic relationships between sovereign nations. Each culture has its own set of intrinsic values and priorities that have varying degrees of emotional ties. It is through these values that people identify themselves, guide their actions, and perceive the world around them. Understanding these values and priorities, the degree of their significance, and their relationship to the dynamics of a region may make the U.S. military more aware of the possible negative consequences of its actions. To be effective in current complex warfare environments, the United States should use its knowledge of other cultures to help establish better working relationships with U.S. allies, civilian populations, enemies, and those who have yet to take sides

However, as the military takes a closer look at employing culture to conduct successful operations in asymmetric environments, it must realize that current training for military personnel is not capable of fully accomplishing the cultural part of its mission. Time, available data and access to skilled cultural experts are limiting factors in the military's ability to include culture effectively in operations. The question then becomes: How does the military access critical information about cultural values and present the information in an easily understandable format?

Ideally, in order to accomplish this goal, a military unit should receive information on the countries and regions it will be entering prior to deploying. This information should come from a variety of sources: country handbooks, the Central Intelligence Agency's country fact book, independent study, professional reading, educational materials, and regional experts. However, if this pre-deployment cultural training plan is adopted, the quantity of information a military unit would receive could be daunting to the average person; for this reason, the military needs to develop a way to filter these data to provide military personnel with only the most essential information. One way the military can filter this information is to use a model to rank the most important aspects of a culture and present the findings to operational units in a graphic, user-friendly way.

This study proposes a framework for comparing cultural values and illustrates its theoretical application for the military by conducting a cultural qualitative analysis of Kuwait—a country that has a non-Western values system. The results of the analysis highlight possible cultural friction points (i.e., cultural values that may cause tension or conflict) between the non-Western-values system country and a Western-values system country. The research suggests that the proposed model is flexible enough to change as information is gathered and refined to a granular level to more specific groups of people (i.e., subregions, cities, ethnic groups, or religious groups). It is only by using a detailed, repeatable methodology the military can accurately determine a baseline value system of any given region; and, in order to be useful, on-scene operational commanders must be able to apply the baseline system in combat.

Culture and Military Operations

When the United States military deploys overseas, whether it is in the capacity of humanitarian relief or counterinsurgency, it is imperative that its members understand the cultural values of the region in order to avoid making strategic errors. In most cases, commanders are left to rely on "country handbooks," or open-source material. Unfortunately, most of this information is not relevant to the commanders' mission, since it does not give the military the tools it needs to track value trends within a specific area of operation. In addition, a proactive commander will need to present the information in a manner specific to the region, subregion, or group of people in his area of responsibility. Once he is able to achieve cultural situational awareness, it must then be transferable laterally as well as vertically to subordinate unit leaders. The objective of cultural studies is to achieve the ability to effectively communicate our message throughout the levels of operation to our target audience.

The Marine Corps *Small Wars Manual*, published in the 1940s, specifically states that Marines must be mindful of three fundamental considerations:

1. Social customs such as class distinctions, dress, and similar items

should be recognized and receive due consideration.

2. Political affiliations or the appearance of political favoritism should be avoided; while a thorough knowledge of the political situation is essential, strict neutrality in such matters should be observed.

3. Respect for religious customs.[1]

These considerations emphasize the need for U.S. military units to attain a familiarity of and respect for the local culture including its language, political and social structure, and economic factors in order to prevent unintentionally creating a hostile environment and methodically to promote "the spirit of good will."[2]

Cultural Values Defined

Cultural values should be significant to anyone who is attempting to interact with a group of people. As Harrison, Lawrence, and Huntington write, "Values matter in how they guide social action. They do so by accounting for the world as it is constructed—making sense of it and why we should even act in it at all in a meaningful way...Values [also] serve different functions for different people."[3] What is perhaps most important in this statement is that values vary greatly among different groups of people, even within the same country or region. For this reason, it is impossible to apply a macro snap shot of culture to the micro or tactical levels of military operation. The larger and more general the scale of the cultural study, the more vague and inaccurate the conclusions or analysis will be. To be effective, cultural studies must be specific to the particular time and place where a military unit is operating.

Cultural Values and Structural Beliefs

Of the various types of cultural values, some researchers such as Coon consider "structural beliefs" (beliefs that form one's identity such as religion, tribal identity etc.) to be the most important.[4] Structural beliefs "usually involve an entity or entities which are related to but are in some senses outside the world in which we live, have at least some improbable and counterintuitive characteristics, and are usually independent of time."[5] In particular, these beliefs are taught to the group's members from a very early age and are absorbed by the members through a process known as imprinting. As a result, these beliefs, or learned behaviors, are difficult to change, unlike learning that occurs later in life. These imprinted beliefs are among those beliefs the group of people hold onto most strongly and, take generations to change or evolve.[6] Furthermore, these beliefs are emotionally binding and are often in the form of ethereal concepts, such as "life after death,".[7]

The premise of emotional bonding or imprinting is that one's belief system directly affects how one sees what the truth is and how one rationalizes a concept. If people are told something that contradicts their beliefs, it is easy to understand why they would view the information with suspicion and come to the conclusion that the information is false.

To suggest that a group of people see the world in the same way as American service members is to assume that the people share the same American beliefs. Regardless of any similarities, it is unlikely that the two groups will share all of the beliefs in the same manner. In fact, a group's belief systems will shape its member's perceptions, and perceptions are more important than what the individual considers to be the truth, especially when one is in a counterinsurgency environment where popular support is the key to success.

Cultural Values and Marketing Messages

Culture and civil conflict are often related. Differences in culture result in different interests, which result in tension and is followed by conflict, violence, and civil war.[8] Only with the assistance of dedicated historians, anthropologists, city planners, and sociologists can the military even begin to understand the complicated and, often, very strange environments in which the military will be tasked to conduct military operations and make meaningful changes.

However, these are not the only entities from which the military can draw; the military can also draw from the business community's best practices to make meaningful changes. For example, when commercial industries market their products overseas, they complete a comprehensive study to aid them in sending the correct message to specific parts of the world. The industries understand that a message that resonates well in the U.S. may not have the same effect in a different culture. As a result, the industries understand that the difference between success and failure is determined in large part by the way they market their product to appeal to the unique perception and values of the local culture. Similarly, the military often has a message it wants to convey to unfamiliar people, and the effectiveness of transferring the message can likewise be equated to operational success or failure.

An example of the commercial sector marketing a message is illustrated in Chrysler's campaign to market the Jeep in France and Germany. The company hired Dr. Clotaire Rapaille, a psychologist, to conduct research and make recommendations to guide Chrysler in effectively marketing its message. In his book, *The Culture Code: An Ingenious Way to Understand Why People Around the World Live and Buy as They Do*,[9] Rapaille develops the concept of a culture code—a system used to identify the values a culture considers most important. He argues that values and beliefs are shaped by one's culture, which gives meaning to objects and actions. Specifically, the code determines three things: 1) how we see ourselves, 2) how others see themselves, and 3) how others see us.

The results of Rapaille's study indicated that the French and Germans could very easily be endeared to the Chrysler Jeep by using historical references. To the French, the Jeep reminded respondents of the American military going into France to end the occupation by Nazi Germany. For the German participants, the Jeep conjured up emotions of the Americans rescuing them from the country's darker days. Based on Rapaille's recommendation, the Jeep, which was formerly called the Wrangler, was renamed the Liberator.

There is no doubt that the Long War, or Global War on Terror, is in essence a war of perception. Tactical and operational decisions often have strategic implications when the media, friendly or adversarial, obtains footage of events and broadcasts the events with their spin. Although the military often fails to effectively spin stories in its favor, the United States is filled with media "spin doctors," as evidenced by the country's election campaigns and commercial marketing abilities. Waging a successful war requires the use of all the elements of national power. In the realm of information, we have yet to use all the resources available to us to shape our message and present it in a fashion that will resonate throughout the various communities of the world and take vital public support away from our enemies.

Military Application of Cultural Intelligence

Although the cultural aspect of operational preparation is now becoming a standard part of war planning, cultural intelligence and awareness is even more vital with respect to counterinsurgency operations in which influence of the masses will ultimately secure victory or result in defeat. Counterinsurgency FM 3-24 expands upon the premise of cultural application by stating that Civil Considerations are an essential part of the Intelligence Preparation of the Battlefield and are "critical to the success of operations."[10] Specifically, the manual identifies six "socio-cultural" factors that should be analyzed to get a clearer picture of the people in a unit's area of operation (AO). Those factors are: society, social structure, culture, language, power, authority, and interests.[11]

Clearly, in each AO some values are more influential than others; the importance placed on each cultural factor may vary, not only from country to country, but even within the same region. This applies not only to specific geographic areas, but to the distinct social groups within the area formed on the basis of tribal, religious, national, and ethnic identity.[12] In order to compare the different importance of various values, this study's proposed model, explained below, assigns relative numerical values to each one of the six "socio-cultural" factors listed in the *Counterinsurgency Manual*.

Even among coalition partners who speak the same language, cultural differences can affect operations and relationships in significant ways. For example, during World War II, British and American soldiers experienced a great deal of friction over perceived cultural insensitivities, such as the way they spoke to each other, or the way they interacted with members of the opposite sex who were of different nationalities. In her 1947 study, "The Application of Anthropological Techniques to Cross National Communication," Dr. Margaret Mead,[13] an American cultural anthropologist, observed cultural insensitivities between British and American soldiers. Specifically, Mead notes that although the British and American

36 *Applications in Operational Culture*

soldiers shared the ideal that the strong are obligated to help the weak, the soldiers developed misconceptions about each other due to more- and less-aggressive methods of communication. The interpretation was that Americans liked to boast and the British were arrogant. Therefore, while the two entities shared similar goals to fight the war and common cultural values, learned habits from childhood translated into communication gaps that resulted in misplaced animosity.[14]

The goal, then, is to evaluate how a member of a cultural group (whether a foreign population or a coalition partner) sees himself, determine what is important to him, and consider it when developing operational objectives, goals, and methods to achieve these operational goals. Specifically, the United States Marine Corps' *Operational Culture for the Warfighter* manual defines operational culture as "those aspects of culture that influence the outcome of

Figure 6

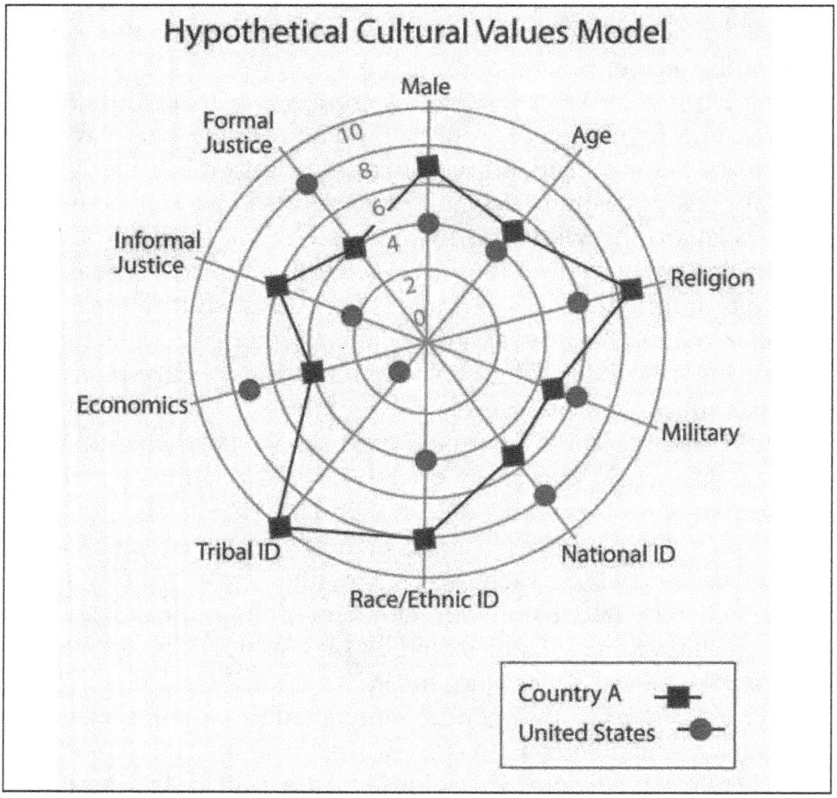

Hypothetical Cultural Values Model

Axes: Male, Age, Religion, Military, National ID, Race/Ethnic ID, Tribal ID, Economics, Informal Justice, Formal Justice

Legend: Country A, United States

a military operation; conversely, the military actions that influence the culture of an area of operation."[15] The fruit of this operational culture research is ultimately to increase the quality and effectiveness of soldiers' communications with competing groups of people in any given area of operation.

A Cultural Values Model

The book *Operational Culture for the Warfighter: Principles and Applications* serves as an excellent starting point for culture studies because it outlines five dimensions through which to view a culture. The dimensions are: physical environment, economic system, social structure, political structure, and beliefs and symbols. Translating these dimensions into the operational situation on the ground, however, requires the ability to filter and focus on the most relevant cultural information. This study proposes a model that can provide a starting point for culture analysis: the Cultural Values Model (CVM). Specifically, the CVM can be leveraged as part of the human terrain preparation for any given region prior to commencing military operations. The CVM can help military personnel by providing a visual, comprehensive representation of the values the local culture finds most important, with the intent of limiting negative interactions caused by cultural misunderstanding and maximizing positive U.S. influence in a given region.

The Cultural Values Model will be introduced, then applied to a case study scenario of Kuwait. To evaluate the case study, the five dimensions suggested by Salmoni and Holmes-Eber will provide the anthropological background and guidance for the analysis.[16] The final product of the application can be used by a commander to define the cultural battlespace during the Marine Corps planning process.

Explanation of the Cultural Values Model

In the proposed Cultural Values Model, cultural values would be ranked (i.e., given a number from one to ten) based on the relative significance the local populace generally places on that value. Thus, the higher the ranking, the more importance a culture places on that particular cultural value. The number would then be placed on the model and each succeeding cultural value would be ranked and the number placed on the diagram to provide a visual image of relative importance between values. In order to ensure accuracy, each cultural value would require two narratives: one that defines the numerical ranking of the value and one that gives a detailed justification for the ranking.

Once the model is completed, an analysis would be conducted to assess the greatest potential for friction points (i.e., cultural values that may cause local tension or conflict between individuals or groups because of the different ranking of importance of the value by the individuals or groups). Furthermore, friction points can be divided into two categories/types: natural and operational.

Natural friction points are any actions that are considered normal by Western standards, but may be offensive to countries with non-Western standards based on the cultural norms of the local populace and might result in friction between U.S. personnel and the local culture. Natural friction points encompass many of the dos and don'ts (food and other taboos, interactions with women or village elders) that are normally briefed as part of a unit's cultural indoctrination; however, using the CVM, operational commanders can determine the friction points pre-deployment or prior to operating in the region by comparing analysis charts to Western cultural values. Specifically, any dimension that demonstrates the greatest difference between U.S. and local cultural values should be considered significant and require special emphasis.

Using the model above, the category of male dominance illustrates an example of natural friction. In this case, male dominance is assigned the highest graded value; and, therefore, is the cultural value Country A finds most important. This means military personnel preparing to deploy to Country A should consider this value to be a friction point. U.S. troops who are accustomed to considering women as peers need to understand that the consequence of misunderstanding or not respecting the cultural value placed on male dominance could result in widespread disdain from the local people.

Operational friction, on the other hand, is the result of military activity and could be intentional or unintentional. Operational friction, once understood and calculated, can be assessed for cultural risk. For example, if a unit commander meets with a local Imam (i.e., an Islamic religious leader) at a mosque, the commander may determine that it is worth the operational risk (or break from operational Standard Operating Procedures) to take his shoes off prior to entering the mosque for the meeting. On the other hand, if the commander and his Marines chase an enemy sniper into a mosque, they would enter the building without hesitation and without removing their footwear. Understanding that it is forbidden to enter a mosque with shoes; the commander should consider apologizing for the intrusion on his way out, in order to convey respect for the religious tradition and reduce the amount of friction incurred. Apologizing, of course, assumes that the Imam was not complicit with the enemy sniper. Showing respect and reestablishing honor sends a very strong message to the religious leader and his followers.

Identifying possible friction points ahead of time allows commanders and unit leaders to make a more focused, condensed model in order to identify specific elements within each category. The Cultural Values Model can be viewed as consisting of numerous layers of specificity. After an initial general cultural assessment has been made, more detailed analyses of key friction points can be undertaken focusing on the details of a specific value area such as religion. For example, in Figure 7, the hypothetical model depicts religion as the most significant value category in an Islamic country. The category of religion is then expanded by creating a separate model with sub-categories (such as holidays, mosques, veils for women) that directly relate to religion in that region.

Similarly, in Figure 7, when the assigned value rankings for the theoretical value of Tribal Identity are compared, the difference

Figure 7

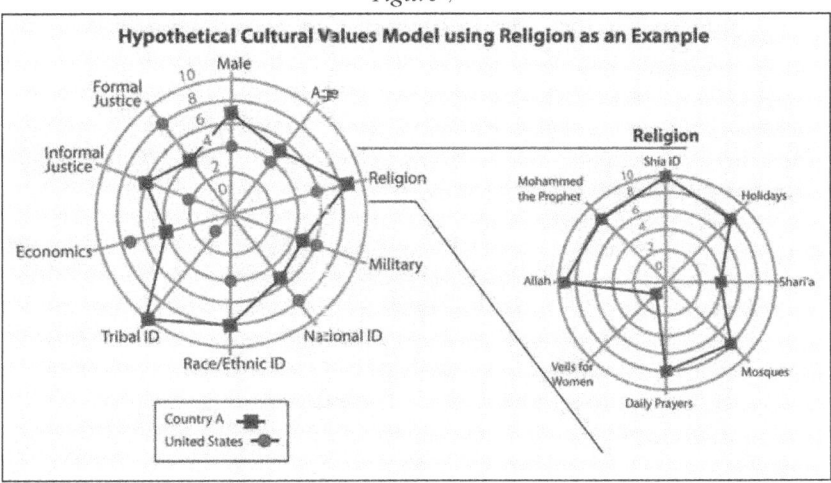

(U.S. = < 2; Country A = 10) between the two rankings indicate an area in which Americans might have the least natural understanding or appreciation of Country A's cultural beliefs. For a commander, this difference would immediately suggest a cultural training requirement for his troops, as lack of preparation for dealing with tribal areas could easily result in unintended friction (i.e., operational friction) with the local population and leadership. The next section provides an illustration of applying the CVM to a case study.

Applications of the Model to a Case Study: Kuwait

National Identification vs. Tribal Identity

The following values model was created from information found in the standard United States Marine Corps' *Cultural Intelligence for Military Operations for Kuwait* produced by the Marine Corps Intelligence Activity.[17] This information is indicative of the type of information that is typically available to an operational unit prior to gaining physical access to a new operational area. Even by limiting the reader to this basic information, one can build a baseline model to brief operational units and to make cross cultural comparisons to other operational environments. The categories and their respective values were determined based on ideas expressed in the Kuwaiti brief and gleaned from an analysis of the ideas, which resulted in common themes about the Kuwaiti people.

Although the information is often qualitative in nature, at some point the analyst must make judgment calls when creating the model. Conclusions that appear to be uncertain should trigger the analyst to conduct additional research in order to improve the level of accuracy of the assessment. In this case, Kuwait was picked because it is geographically small and the population is relatively homogeneous in comparison to most other countries in the world (See Figure 8).

Figure 8 provides the foundation for a briefing to explain what is important in the Kuwaiti Operational Area and how the Kuwaitis' main value points compare to those of Americans. An important point to note when examining the Kuwait model is that National Identification is the most significant aspect of culture for Kuwaitis. This point is significant for two reasons: First, national identification

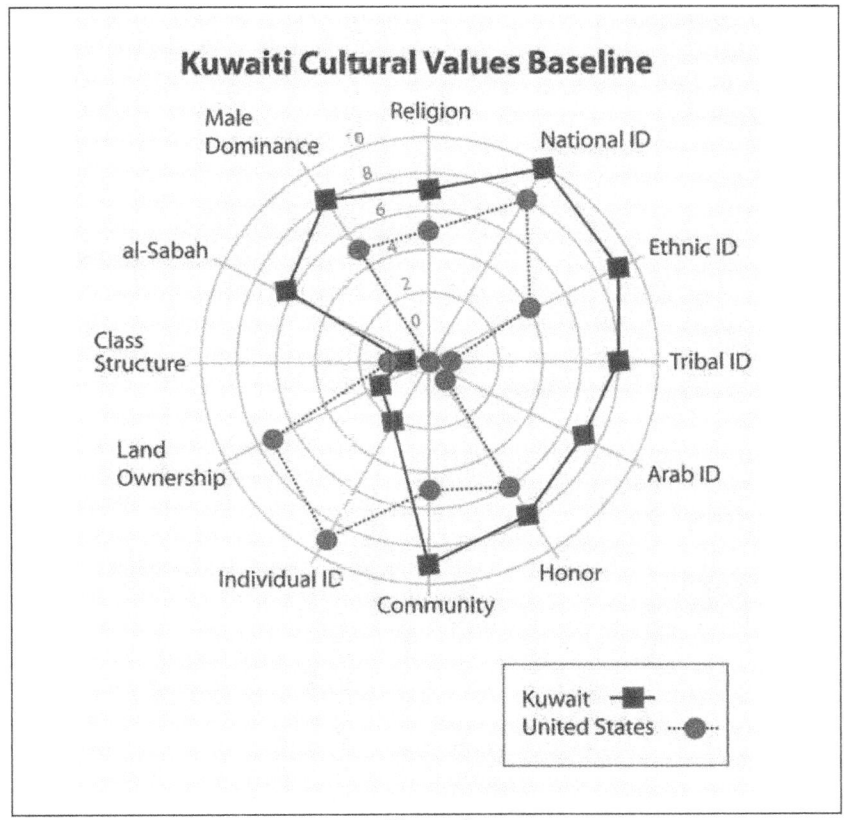

Figure 8

has only recently become a reality in the culture because prior to the 1991 invasion of Kuwait by Iraq, tribal identity was typically more significant to the people than was national identity. However, the 1991 invasion of Kuwait, coupled with an active program emphasizing national identity on the part of the ruling al-Sabah family, had a galvanizing effect on the Kuwaitis. Now, national identification supersedes tribal affiliations and loyalty to merchant families.[18]

Kuwaiti Citizens vs. Foreign Nationals

In addition, the CVM depicts the importance Kuwaitis place on various groups of people. For example, the societal hierarchy that emerged when the graph in Figure 9 below was created shows that Male Kuwaiti Citizens are at the top of the list, followed by Female

42   *Applications in Operational Culture*

Kuwaiti Citizens, Arabs, Others—Skilled Foreign Workers, and, at the bottom, Others—Unskilled Foreign Workers.[19] This finding is significant because a major source of tension in Kuwait is with its overwhelmingly large group of foreign workers. Foreigners are not held in very high regard and, although they are transient, they make up about 50 percent of Kuwait's population.[20] The model clearly shows why there would be tension (i.e., natural friction) in this category. Yet, interestingly enough, with the emphasis on Kuwaiti national identity, this potential area of friction appears moderated as a result.

Land Ownership vs. Trespassing

Another area that westerners might find difficult to understand without additional research is the concept of trespassing (i.e., a subcategory of land ownership). Based on Bedouin traditions, Kuwaitis have very little regard for respecting personal property lines. As a result, Kuwaitis do not readily see an issue with people walking across property that does not belong to them.[21] Em-

Figure 9

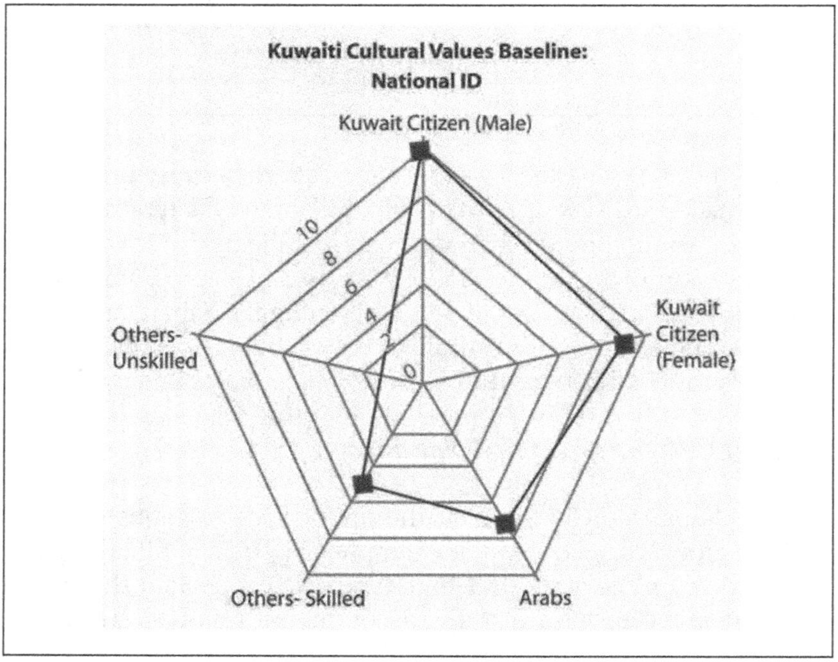

Kuwaiti Cultural Values Baseline: National ID

ployees of a foreign company operating in Kuwait could easily draw negative attention if they did not understand this cultural belief and tried to enforce a different set of norms by refusing to allow Kuwaitis to walk on their land.

Regionally, Kuwait is unique in many historical ways, which in turn makes its cultural values somewhat distinctive to surrounding Middle Eastern nations. A case in point is that because Kuwait was never subjected to colonialism, the local people do not feel the sense of social oppression that seems to endure among people whose nations were subjected to colonialism. Specifically, a CVM of Iran would show that unlike Kuwait, Iran's national identity is very strongly connected to its historical lineage to "the great Persian Empire."[22]

Kuwait vs. U.S. Cultural Values

When examined in terms of cross-cultural comparison with the United States, one can see distinctions between Kuwait and the U.S. that are worth noting For example, Kuwait is a male dominated society. Although the argument can be made that the U.S. is also male dominated, the degree is far less than in Kuwait. There are also vast differences between Kuwait's al-Sabah family and tribal structure and America's individualistic culture. In cases such as this one, a commander could use the Cultural Values Model as an indicator of cultural gaps; the culture briefing military personnel receive would then be able to adequately prepare them for the types of cultural friction—natural and operational—that they will experience.

Discussion of Findings

The Cultural Values Model provides a baseline of information that is open to refinement and expansion as more information is gathered, or as experts add their perspectives. The CVM could be applied by starting with a national or regional level model, then refining this model to the micro level as needed to cover different groups of people or regions. For example, in larger, less homogeneous countries like Iran, a model would need to be created for the country as a whole, then for the major cities, and rural areas. A national-level model would reveal the significance of its historical heritage. As the model was refined to smaller regions, the importance of those regions' histories would continue to be a significant factor.[23] In contrast, when examining the role of women, one would see indications in Tehran that women are afforded the opportunity to achieve prominence and share a great deal of equality with men that is unmatched throughout most of the rest of the Middle East. The reason for this is that Tehran is considered to have been more influenced by westernized values than the rural parts of Iran where traditional male dominated norms have been maintained.[24]

Regional research has indicated that there are diverse cultural groups with dissimilar values in areas that we might have otherwise expected to be more homogenous. Specifically in Iran, the population could be categorized along the various geographic, ethnic, religious, linguistic, or tribal lines because each category clearly shows the magnitude of the diversity of the Iranian population.[25] As the models are refined for smaller geographic regions or groups of people, they will increase in accuracy.

Conclusion

Cultural understanding is more than knowing the dos and don'ts of a given region. Cultural understanding occurs when a person is able to grasp how someone else perceives the world around him or her and that person can anticipate the other's reaction to events. The only way to reach that point of cultural enlightenment is through intense study and assistance from professionals who have conducted research in the area. It is unlikely that very many military leaders will have the opportunity to reach the point of regional expertise combined with a PhD-level education in a professional discipline such as cultural anthropology. Regardless of the circumstances, the U.S. military, and in particular the U.S. Marine Corps, has an obligation to build its organizational structure to institutionalize culture as a formal line of operation.

As military operations continue to focus on the importance of winning "hearts and minds," cultural understanding has become increasingly important. For the average Marine, a user-friendly model that represents the values of a region would help him or her to attain the necessary cultural understanding to influence the people of a particular region. Therefore, it is important for the model to be flexible, simple, and comprehensive. However, in order for this flexible, simplistic, and comprehensive model to be functional, it must also be accurate. If the input data are flawed, regardless of how it is analyzed or processed, the output will also be flawed.

To improve accuracy and relevance, anthropologists and regional experts should be included in the development and evaluation of the model. For this reason, the model should be developed and maintained outside of the Intelligence section, so that professional participation by regional experts and anthropologists would not be placed in a situation that would put them at ethical odds with their professional ethos. The only way to properly refine this type of model and ensure its accuracy is to use regional experts in its development. The initial template would include major items for consideration, but it would not be restricted to those specific items.

For the military, cultural values and differences can be significant factors in influencing conflict resolution. To neglect culture when attempting to resolve violence makes any resolution unlikely to have long term success.[26] Cultural studies will enhance the U.S. military's communication abilities, not only on the operational level, but also on the tactical level. From tactical decision making to information operations, to strategic communications, detailed cultural understanding plays a significant role in the success or failure of an operation. The U.S. government and military has the ability to leverage a large pool of civilian talent to market its message to communities worldwide and to tailor those messages to specific regions of political interest. In the War on Terror/Long War, the U.S. should be able to do a better job marketing its position and world views to Middle Eastern regions in a manner that resonates with their cultural values and perceptions.

Operational culture is an aspect of planning that will not be short lived. For the foreseeable future, cultural values assessments, coupled with their application to information operations, will play an integral role in all military operations. However, agreement on the actual process will likely involve a lengthy debate. Even within the anthropology community, there are many methods and disagreements concerning operational culture. Like the model in this paper, field work or operational experience is the key to creating an accurate model.[27]

Still in its infancy, this integration of academic studies (i.e., cross cultural modeling) with information and counterinsurgency operations will require a great deal of maturing. Ideally, one day the concept will become as important as operations, logistics, and intelligence. Even more importantly, this discipline, hopefully, will form the nucleus for interagency operations, guiding how the U.S. government allocates its precious resources to achieve holistic, long-term solutions to complicated problems.

Notes

[1] U.S. Marine Corps, *Small Wars Manual* (Washington, DC: U.S. Marine Corps, 1940), 19.

[2] *Small Wars Manual*, 45.

[3] Lawrence Harrison and Samuel Huntington, *Culture Matters: How Values Shape Human Progress* (New York: Basic Books, 2000), 148.

[4] Carleton S. Coon, *Culture Wars and the Global Village: A Diplomat's Perspective* (Buffalo, NY: Prometheus Books, 2000).

[5] Ibid., 102.

[6] Ibid.

[7] Ibid., 108.

[8] Bettina Schmidt and Schroder Ingo, *Anthropology of Violence and Conflict* (New York: Routledge, 2001), 179.

[9] Clotaire Rapaille, *The Culture Code: An Ingenious Way to Understand Why People Around the World Live and Buy as They Do* (New York: Broadway Books, 2006), 3.

[10] HQ Department of the Army, Washington, DC; HQ Marine Corps Combat Development Command, Department of the Navy; HQ United States Marine Corps, Washington, DC, *FM 3-24 (MCWP 3-33.5) Counterinsurgency* (Washington, DC: U.S. Department of Defense, 2006), section 3-3.

[11] Ibid., 3-3 to 3-4.

[12] Ibid., 3-4 to 3-5.

[13] Margaret Mead, "The Application of Anthropological Techniques to Cross National Communication," *Transactions of the New York Academy of Sciences*, ser. 2, vol. 9, no. 4 (1947): 133-52.

[14] Ibid., 107-10.

[15] Barak A. Salmoni and Paula Holmes-Eber, *Operational Culture for the Warfighter* (Quantico, VA: Marine Corps University Press, 2008), 14.

[16] Ibid., 51.

[17] *Cultural Intelligence for Military Operations: Kuwait* (Quantico, VA: Marine Corps Intelligence Activity, 2003).

[18] Ibid.

[19] Ibid.

[20] Ibid.

[21] Ibid..

[22] Massoume Price, *Iran's Diverse Peoples: A Reference Sourcebook* (Santa Barbara, CA: ABC-CLIO, 2005), 267.

[23] Encyclopedia Britannica, *Iran: The Essential Guide to a Country on the Brink* (New York: Wiley, 2006), 45.

[24] Helen Chapin Metz, *Iran: A Country Study*, 4th ed. (Washington, DC: U.S. Government Printing Office, 1989).

[25] Price, *Iran's Diverse Peoples*, 92-131.

[26] James Bradford, *The Military and Conflict Between Cultures: Soldiers at the Interface* (College Station: Texas A&M University Press, 1997), 205-11.

[27] Frank W. Moore, *Readings in Cross-Cultural Methodology* (New Haven, CT: Human Relations Area Files, 1966), 50-76.

Chapter 3

Developing the Iraqi Army: The Long Fight in the Long War

Major John E. Bilas, USMC

Major John E. Bilas graduated from Youngstown State University in 1987 with a degree in applied science. He was commissioned a second lieutenant in August 1987 and served with 3d Battalion, 8th Marine Regiment. Following a tour at Camp Lejeune, North Carolina, Bilas served at Marine Barracks, Washington D.C. (8th and I), where he was a ceremonial platoon commander, silent drill team platoon commander, and assistant operations officer. Following his tour, Bilas was released from active duty in 1994 and had a successful career in business, rising to general manager for a Fortune 500 company.

After the events of 11 September 2001, Bilas returned to the Marine Corps, and his commission was reinstated in May 2002. He was stationed with 2d Marine Division at Camp Lejeune and served as executive officer and company commander for Small Craft Company. He then made a lateral move into intelligence, and after attending the MAGTF Intelligence Course in Dam Neck, Virginia, in 2004, Bilas transferred to Camp Pendleton, California, and served as assistant operations officer for the MEF G-2, intelligence advisor for an Iraqi brigade, and future operations planner for the MEF G-3. Bilas attended Command and Staff College at Marine Corps University and received his Masters of Military Studies in June 2008. He is currently the S-2 for TTECG at Mojave Viper, Twentynine Palms, California. Bilas has received the Navy Marine Corps Achievement Medal, the Navy Marine Corps Commendation Medal, and the Bronze Star.

The strategic decisions leading up to the Iraq War will be discussed and debated for many years. After the fall of the Iraqi regime, American forces were not prepared for, nor had they planned for post-combat operations. While many factors led to U.S.

challenges in post combat stability operations, one factor in particular—the strategic decision making regarding the disbandment of the Iraqi Army—has been a central point of debate within the U.S. military.

In 2003, the Iraqi Army was disbanded and Coalition Provisional Authority (CPA) Order Number 2, 23 August 2003,[1] dictated a new army was to be formed. Inadequate long-range strategic design, the disbanding of Iraqi Security Forces, and the lack of understanding of the Middle-Eastern culture cost not only American lives, but also left Iraq in lawlessness and chaos. As a result, the U.S. Military was forced to play a larger role in nation building—in particular, building a new Iraqi Army.

The process of building a new Iraqi Army has been far from simple. Beginning in 2004, inadequate logistics, personnel recruitment and retention, and pay administration within the Iraqi Security Forces (ISF) were major concerns for Coalition forces. Furthermore, progress toward independence was impeded at top levels within the Government of Iraq (GOI) and the Ministry of Defense (MOD) because of poor security conditions and continuous corruption within the ministry. Despite these challenges however, there are indications of progress in developing the Iraqi Army in the past three years.

Using a case study approach based on observational field data collected during 2006 and 2007, this chapter will evaluate successes and failures in U.S. efforts to rebuild the Iraqi Army, its development, and its ability to sustain itself in combat service support functions. Based on the data, the author argues that the surge of American troops, coupled with local and militia uprisings (i.e., the al-Anbar Awakening), were catalysts for the Iraqi Army's (IA) progress in critical areas, such as logistics, personnel recruitment and retention, and pay administration, which contributed to building the confidence and performance of the IA in 2007. In addition, the analysis illustrates that progress at the tactical level has outpaced progress within the GOI's ministries at the strategic level.

Methodology

This paper uses an explanatory case study approach to investigate the 2d Brigade, 7th IA Division's (2/7 BDE) level of progress

Developing the Iraqi Army 51

made in logistics, personnel recruitment, and pay administration over the span of two calendar years, 2006 (CY06) and 2007 (CY07). The case study approach "involves extensive observation of a single individual, several individuals, or a single group of individuals as a unit"[2] in order to determine the relationship among the various factors that influence the current behavior of the 2/7 BDE. An explanatory case study concentrates on "experiential knowledge of the individuals or unit and close attention to the influence of their/its social, political, and other contexts."[3]

In order to determine the relationship among the brigade's various factors, several sources were used for data. For the experiential knowledge, in 2006 the researcher served as assistant operations officer for the MEF G-2, intelligence advisor for an Iraqi brigade, and future operations planner for the MEF G-3. The researcher also interviewed a former advisor of 2/7 BDE, located in Al Asad, Iraq. He was the researcher's successor upon turnover of the researcher's duties in December 2006. The researcher's personal experience in 2006 will be compared to the successor's assessment of the BDE's performance in 2007 in the al-Anbar region. The researcher's personal notes and successor interview are the primary sources.

Similarly, U.S. advisor after action reports (AAR) regarding other Iraqi units were reviewed. Additionally, the researcher wanted to gain perspective from outside of al-Anbar through a personal interview of a former U.S. Army advisor, who deployed with the Iraqi Army's 2d Division in Mosul. In addition to other advisors' reports, the researcher evaluated Government Accountability Office (GAO) reports, *The Report of the Independent Commission on the Security of Iraq, Report to Congress on the Situation in Iraq*, and other newsprint articles (e.g., newspapers, *Marine Corps Gazette*) in order to look at factors that affected the Brigade, to include Iraqi units outside of al-Anbar, and analyze how those factors were influenced by the Brigade's social, political, and geographic contexts.

By using a wide variety of resources (i.e., government documents, first-person experiences and observations, personal interviews with subject matter experts), the researcher was able to triangulate the data, which highlighted common supporting themes from several sources. An Excel spreadsheet was used to categorize large amounts of data into its respective categories and to cap-

52 *Applications in Operational Culture*

ture specific information for each area. An example of how the researcher reduced the data for each category (logistics and pay) can be seen in Appendix A and B. The following section introduces how the surge and civilian uprisings improved the performance of the Iraqi BDE.

Background Information

A second tenet of counterinsurgency is that foreign forces, however adept, are no substitute for capable indigenous ones. Lest we forget, huge foreign military intervention did not produce victory for the French in Indochina or Algeria, for the Soviets in Afghanistan, or for the Americans in Vietnam. An indigenous army can claim a measure of legitimacy that an occupying army—especially a Western one amid Muslim cooperation—cannot.[4]

In 2003, after the invasion of Iraq, Coalition forces misjudged the character and nature of the conflict and failed to plan for post-conflict and reconstruction operations. That is, Coalition forces transitioned from liberators to occupiers. In addition, the decision to disband the Iraqi Army in May 2003 was also a strategic flaw. The outcome of this decision resonates as one of the most failed decisions during the war because disbanding the Iraqi Army only exacerbated the security situation. In an article from the *The Interagency and Counterinsurgency Warfare,* James J. Wirtz wrote "The Exquisite Problem of Victory: Measuring Success in Unconventional Operations." Wirtz states:

> Greatly worsening the situation was the U.S. decision to disestablish the Iraqi Army, a decision that promptly left over 400,000 officers and soldiers unemployed. This military talent served as a ready recruitment pool for ethnic militias, criminal organizations, and terrorist groups who benefited from the chaotic conditions that quickly spread across several urban centers in Iraq...on stability operations, the Coalition's efforts in Iraq were more than unsuccessful. They actually were counterproductive because they had the net effect of reducing the security enjoyed by the Iraqi people.[5]

Effects of Operation Fardh Al-Qanoon on Iraqi Provinces

PROVINCE	DEVELOPMENT
ANBAR	* Violent attacks in the Ramadi region have dropped from 25 per day in 2006 to 4 per day since the Surge (April 29, 2007) * In May 2006, There were 811 attacks throughout the province. In May 2007, that figure was just over 400 (May 31, 2007) - In the city of Ramadi, there were 234 attacks in May 2006 compared to 30 in May 2007 * Since the beginning of 2007, 12,000 Iraqis have volunteered for the security forces. In all of 2006, 1,000 volunteered (May 31, 2007)

DIYALA	* There has been roughly a 30% increase in offensive actions and attacks in Diyala province (March 9, 2007)
	
	* In 2006, Diyala province was the eighth-deadliest province (of Iraq's 18) for U.S. troops (April 22, 2007)
	   - Thus far in 2007, it ranks as the third-deadliest province behind Baghdad and Anbar
	
	* Over the past five months, attacks on U.S. and Iraqi troops have increased 70% (April 16, 2007)
	   - It was reported on April 15, 2007, that almost a full brigade of between 2,000 and 3,000 soldiers is being sent to reinforce the territory between Baghdad and Baqubah, the provincial capital.

BAGHDAD	In all of 2006, 266 weapons caches were found within all security districts. Thus far in 2007, 411 have been found (May 31, 2007)

Figure 10

Source: Iraq Index, Tracking Variables of Reconstruction and Security in Post Saddam Iraq, The Brookings Institution, 21 December 2007, 15.

By July and August 2003, the first efforts were taken to form and create the new Iraqi Army. These efforts were led by the Coalition Military Assistance Training Team (CMATT). CMATT planned to establish three light motorized brigades, called the Iraqi Civil De-

54   *Applications in Operational Culture*

fense Corps (ICDC). However, structural and organizational problems within CMATT and the CPA hindered the effort to operationalize these brigades.[6] As a result of the failure to accurately plan for and resource the Iraqi military, training and operations were further delayed. It was not until the spring of 2004 that Coalition forces achieved a sense of urgency and realized Iraqi security forces were vital to Iraq's future.[7]

Between 2004 and 2005, the Iraqi Army continued to struggle. Operational capability was limited and personnel attrition remained extremely high. As Coalition forces continued to form, train, man, and equip a new Iraqi Army, security in Iraq continued to deteriorate, which put a continuous strain on Coalition forces. Furthermore, many Iraqi units performed abysmally. Iraqi units were poorly led, refused to fight, and deserted from the army.[8] According to Cordesman, "Corruption, nepotism, political favoritism, false manpower reports, and false activity reports continued to be serious problems—something that was inevitable in trying to develop security and police forces in a foreign culture and in a country that was governed in a corrupt manner for decades."[9] To stanch this poor performance, the U.S. placed advisors with Iraqi Army units. The following section will describe how transition teams were formed to advise the new Iraqi Army in order to help bring security and order to the country.

The Formation of Military Transition Teams

While Coalition forces were providing security throughout Iraq, advisor teams were created to help develop the Iraqi Security Forces (ISF). Specifically for the IA, the Military Transition Teams (MiTTs) were embedded with Iraqi units for training, mentoring, and advising. Advisor teams were task-organized to ensure team composition facilitated the Iraqi Army in the functional areas of operations, intelligence, logistics, communications, and leadership. However, in 2006, many advisors were redirected to help build the Iraqi Army's life support infrastructure, instead of training them to fight a counterinsurgency. The 2/7BDE's operational status during the time that the researcher was assigned to it was no different. His team was not only tasked with building the BDE's fighting capability, but was also part of the effort to build the BDE's life support infrastructure.

The researcher is an intelligence officer by training, and his expectation for his tour of duty in Iraq in 2006 was to train and mentor an Iraqi BDE Intelligence Section. Instead, his mission was redirected toward building the foundation for the Iraqi BDE. Able to conduct limited combat operations, the BDE was incapable of sustaining any level of progress because the basic life support requirements of logistics, personnel, and a pay system was broken. This research paper is grounded in the researcher's day-to-day experiences and interactions with members of the BDE and MOD.

The security situation during 2006 was poor and hindered the researcher and his team's effort to establish the BDE's operational capability; this problem was also pandemic across all Iraqi units. As a result, General David H. Petraeus recommended that 30,000 additional troops deploy to Iraq in December 2006. The researcher was interested to find out how the BDE functioned as a result of the improved security, and this paper's findings show that the improved security had a positive impact on the BDE's progress.

The Surge—Operation Fardh-Al Qanoon and the Awakening

By December 2006, Coalition forces were failing to meet their objectives in Iraq. The peak of the sectarian violence so far was the bombing of the Golden Dome Mosque in Samarra in February 2006, which escalated the level of violence throughout the country.[10] There were not enough U.S. or Coalition troops to secure the country, and the level of violence hindered the progress of the GOI and the Iraqi Army.[11] Because the country was in complete shamble, the IA was incapable of conducting independent operations and was totally dependent on Coalition forces for survival. All combat service and support was provided by Coalition forces. Morale was low within the Iraqi Army and soldiers were quitting in droves.[12] By mid-June 2007, 30,000 new U.S. troops were deployed throughout the country. In al-Anbar, a Marine Expeditionary Unit (MEU) and one infantry battalion deployed to conduct surge operations, which totaled nearly 5,000 additional Marines and Sailors.

General David H. Petraeus, Commander of Coalition forces in Iraq, focused the surge on the following areas:

- Perform offensive and clearing operations in several key cities.

- Engage in dialogue with insurgent groups and tribes to oppose al-Qaeda (AQI) and other extremists.
- Emphasize the development of the Iraqi Security Forces.
- Employ non-kinetic operations with added reconstruction teams.[13]

Once offensive operations commenced, al-Qaeda in Iraq's (AQI) influence and ethno-sectarian violence was reduced.[14] In addition, an uprising of anti-al-Qaeda groups formed under the leadership of various Sheiks in al-Anbar. A plethora of sources confirm that the violence and uprisings were beginning to decline. According to an article written in the *Weekly Standard* in March 2007, the "awakening" was a key to reducing the level of violence.[15] On 28 December 2007, *Time* magazine reported that "militias that once fought against Coalition forces have turned against al-Qaeda."[16] In addition, on 21 December 2007, the *Washington Post* reported, "the number of attacks fell in al-Anbar from 1,350 in October 2006 to fewer than 100 per month. Last week, there were just 12 attacks in Anbar."[17] Figure 10 depicts a portion of the effects of the surge.

Furthermore, the Concerned Local Citizens (CLC) program was instituted and provided local security in Baghdad. Recruited citizens were paid $300 to join in neighborhood watch groups. As of December 2007, there were 69,000 citizens serving in this program.[18] Finally, on December 30, 2007, the *Washington Post* reported:

> The downturn in violence is generally attributed to three factors that emerged over the year: the arrival of 30,000 additional troops, the emergence of tens of thousands of Sunni fighters who aligned with American troops against al-Qaeda in Iraq, and the decision by Shiite cleric Moqtada al-Sadr to call for a six-month cease-fire by his militia.[19]

For the IA, the surge improved the security and stability for cities and road networks to support logistical and personnel movement. In addition, the al-Anbar Awakening and the CLC provided Coalition forces useful tactical intelligence to target insurgent activities and AQI. Therefore, these initiatives extended the space and time available for Coalition forces to operationally develop the IA and sustain themselves independently. Because of these events, the IA has made considerable progress. David Gompert, an analyst from the Rand Corporation, noted:

> The army is emerging as one of Iraq's few effective national institutions: an integrated force with units and commanders drawn from every province, people, and sect. Generally speaking, they are well-led, disciplined, adequately funded, politically trustworthy, and respected by the people. The most important indicator of their progress is that U.S. commanders are starting to decide that Iraqi Brigades can replace U.S. brigades one by one, a process that will gain speed in the coming year.[20]

The evidence presented above indicates how the surge of personnel provided Coalition forces, specifically the MiTTs, the space and time to continue the development of the IA. Furthermore, it is important to mention that by summer of 2008, the surge allowed the Iraqi Army to stand on their own to operate and sustain themselves independently without Coalition forces' assistance. However, it is expected that advisors will still embed with the IA, but operations, intelligence, logistics, and personnel/pay administration will be led by the Iraqi Army.

In 2007, 2/7 BDE made significant improvements in their operational development, in spite of MOD's slow progress to develop and implement its policies to support the IA; and, in fact, Major Steve Sims, 2/7 BDE MiTT in 2007 (the researcher's successor), stated that the BDE was operating independently and had tactical control of assigned battlespace. During the interview, Sims noted that 2/7 BDE was currently rated as Operational Readiness Assessment (ORA) Level 2 and had reached that level within six to eight months at the end of 2006.[21] (The BDE had tactical control over its battalions and the 7th Division had tactical control over its Brigades.) (Figure 11 provides a description of the ORA).

In addition, Iraqi Ground Forces Command (IGFC) had tactical control over the two Divisions in al-Anbar. In essence, 2/7 BDE was operating in battlespace adjacent to 2d Regimental Combat

Iraqi Army Operational Readiness Assessment (ORA) Level Definitions.

Green	A Level 1 unit is capable of planning, executing, and sustaining counterinsurgency operations.
Yellow	A Level 2 unit is capable of planning, executing, and sustaining counterinsurgency operations with ISF or coalition support.
Orange	A Level 3 unit is partially capable of conducting counterinsurgency operations in conjunction with coalition units.
Red	A Level 4 unit is forming and/or incapable of conducting counterinsurgency operations.

Figure 11

Team (RCT-2) and all tactical direction and tasks were now directed from the Iraqi Division. Sims stated that the BDE was "performing well in tactical and convoy operations." Sims continued:

> The BDE's Operations Shop grew by two field grade officers. A civil affairs and training section was added to the BDE staff. Planning continued to remain centralized and the BDE Commander still made all the decisions. Even though the staff was taught numerous classes on Military Decision Making (MDMP), the BDE staff always reverted back to the BDE commander to make all the decisions. Nevertheless, the BDE executed well. The soldiers performed well in tactical operations. They [Iraqi soldiers] still lacked discipline for not wearing their PPE [Personal Protective Equipment], but they still got the job done.[22]

This analysis illustrates that the BDE's overall operational capability improved over the course of the year. Yet, according to the researcher's and his successor's personal observations, it was the surge, coupled with the rising of the Iraqi people, that most impacted the critical areas of logistics, personnel, and pay.

How the Surge Affected Logistics, Personnel, and Pay

Logistics

In war, a military unit's capability is degraded if the unit has poor logistics planning and execution.[23] During the researcher's tour of duty, he, along with his team members, determined that IA logistics was the bottleneck to effectively enhance the BDE's sustainment capability.

A logistics concept was developed by MOD in 2005, as a multi-layered system that synchronized the logistics and maintenance at all levels (i.e., strategic, operational, tactical), within the Iraqi Army.[24] As a result, various garrison support units and maintenance contracts were established in key locations around Iraq. Strategically, MOD developed a support command that was responsible for plans, policies, acquisition, and budgeting. Formal schools were also established to train logistic specialists supported by MOD and Coalition forces. The schools, located in Tajji, provided a formal curriculum that enabled officers and enlisted personnel to acquire the capability and skill level to support the logistical infrastructure within the army.[25]

Logistics: 2006

Unfortunately, MOD was unable to fully execute and implement this logistics plan. Security conditions and MOD's inefficiency and inexperience slowly hampered the logistics development in 2005 and 2006, although some progress was made in 2007.[26] In addition, in September 2007, The Independent Commission reported, "The lack of logistics experience and expertise within the Iraqi armed forces is substantial and hampers their readiness and capability. Renewed emphasis on Coalition mentoring and technical support will be required to remedy this condition."[27] For 2/7 BDE, the Team Chief stated in June 2006, "Sometimes they don't eat for days....logistics has been the Iraqi army's primary problem here."[28] Considered the Achilles heel of developing the IA, logistics has had uneven progress in 2007, as noted in a recent Department of Defense Report, *Measuring and Stability in Iraq*:

> The MoD, and to a lesser extent, the MoI, have

shown some improvement in logistics capabilities. The notable exception is an inability to adequately forecast life-support requirements and to promptly take action when contracts are expiring. The Minister of Defense had set an ambitious goal of December 1, 2007, for attempting again to assume life support self-reliance. Multi-National Security Transition Command-Iraq (MNSTC-I) advisors had recommended that implementation be phased in over time. So far, the implementation has been mixed. The construction of national-level maintenance and warehousing facilities at the Taji National Maintenance and Supply Depots should be completed by 2009.[29]

Furthermore, Coalition forces instituted a type of "tough love" policy toward the Iraqi Army by mid-2006. Because Coalition forces had a time-driven plan for IA logistics independence, this policy halted all Coalition forces logistic support including fuel replenishment to the IA. This "kick-start" initiative allowed the IA to begin demonstrating their logistical capability in order to gain operational independence. This policy failed because the IA/MOD infrastructure was still not established and the security conditions were still not favorable for independent Iraqi convoys for logistical movement. Coalition forces then resumed combat support to the IA and realized that the IA's logistics transition was going to be an event-driven process based on improved security conditions as opposed to a timeline schedule.[30]

Logistics: 2007

2007 was considered the "Year of Leaders and Logistics"[31] as stated by MNSTC-I's campaign plan. The plan stated that the Iraqi Joint Headquarters was to assume total responsibility for logistics by November 2007. This benchmark was not totally achieved; however, considerable progress was made within the BDE.

In January 2007, MOD awarded the Sandi Group (Iraqi food service contractors with U.S. oversight) the contract to oversee the food and life support for the IA, and significant progress was made in BDE's quality of life.[32] Food distribution improved because the surge increased the road networks' security. In addition, the Sandi Group was also able to support fuel transportation across al-Anbar, and by mid 2007, the entire BDE's fuel requirements became independent from the Coalition forces. Furthermore, the road networks improved to the point that Iraqi commanders did not want American convoys and checkpoints on the main roads because it caused traffic delays for Iraqi logistical resupplies.³³

Personnel

Manpower retention and recruitment were also major obstacles for MiTTs and the 2/7 BDE to overcome. These personnel shortfalls also inhibited the IA's operational capability. Because Iraq was still unsecure as of 2002, recruitment and sustainment of young Iraqi males was difficult to achieve. Between 2006 and 2007, however, a significant shift occurred in the ability of the IA to recruit and retain personnel.

Personnel: 2006

In June 2006, the *Stars and Stripes* reported, "In Hadithah, the Iraqi Army Brigade has been losing about 100 soldiers a month, dropping from more than 2,000 at the beginning of the year to fewer than 1,600 in May."³⁴ The personnel decline was also noted in the same *Stars and Stripes* article by the researcher's MiTT Chief. "We won't make any real progress until we stop hemorrhaging the personnel," stated Lieutenant Colonel Jeffrey J. Kenney.³⁵ 2/7 Brigade was not the only unit experiencing high levels of attrition. In an article referring to Iraqi units across al-Anbar, Dr. Carter Malkasian, from the Center of Naval Analysis, wrote:

> Between 20 and 33 percent of the 750 men in a battalion are on leave at any time, while desertions and combat losses—because of poor living conditions, irregular pay, distance from home, and constant exposure to combat—reduced on-hand strength to between 150 and 600 men per battalion. In the worst cases, personnel attrition has forced certain Iraqi units to drastically cut back on operations.³⁶

Recruiting drives proved unsuccessful in 2006 due to several factors. The BDE commander had a positive influence with the local Sheiks, but was unable to fully persuade sufficient number of Sunni males to join the predominantly Shia Army. The sheikhs affirmed that no males were going to join the army because of their fear of becoming targets of the insurgents.[37] The BDE and Coalition forces were unable to win the support of the local Sunni community because of the strong influence AQI had on the local population. It was clear that local support for police, army, and the government was not going to happen until the insurgency was quelled.[38]

Personnel: 2007

Fortunately, there was a turning point in 2007. As the surge began to eliminate AQI's influence, especially in al-Anbar, the local sheikhs also began to sway toward local and regional reconciliation. The al-Anbar Awakening was the "call to all of Iraq to stand up with people from all tribes, and all religions to stop insurgents from causing fear throughout their land."[39] In addition, Major General Walter E. Gaskin, Commanding General, Multi National Force-West, stated, "Here around us is the evidence of the peace and stability that are the rewards of the Iraqi people standing together, standing to eliminate the terrorism and lawlessness."[40]

In addition to the BDE Commander's growing influence, sheikh persuasion, pay incentives, and quality of life were attractions that led Sunni males to join the army. During 2007, the BDE held six recruiting drives from Rawah, Hadithah, Baghdad, and Hit and was able to recruit over 6,300 Sunni males into the IA. This integration of Sunni and Shia soldiers into the army also had an impact on the leave process. Figures 12 and 13 depict the growth in personnel in 2/7 BDE between 2006 and 2007.

Modified Table of Equipment (MTOE) = The Iraqi Army's manning and equipment organization. The MTOE was developed in 2005 based on Coalition Provisional Authority (CPA) Order Number 2, which disbanded the old IA and created a new structure based on coalition input.

The Shia soldiers continued to travel by bus from Al Asad to Baghdad for their monthly leave period.[41] Because security improved over the course of the year, bus transportation traveled

Personnel Strength for 2d Brigade, 7th Iraqi Division (2006-2007)

	On Hand	Present for Duty	Fired per Month	Wounded in Action	Killed in Action
2006	850	600	100	170	35
2007	2,800	1,600	40	50	15

Figure 12

Source: Compilation of numbers from Bilas' Personal Journal, 2006, and from Sims' personal interview on 11 December 2007.

more freely between Baghdad and Al Asad than in 2006. In addition, the BDE was providing the security for the passenger buses—a task that Coalition forces performed during 2006.[42] The Sunnis, however, no longer required bus transportation and were able to freely depart Al Asad Base on leave and travel to their homes within the local area.[43]

Personnel strength also grew across the entire Division throughout al-Anbar. By late 2007 the BDE was able to reach its personnel manpower goals according to the Modified Tables of Organization and Equipment (MTOE), which is the IA's task organization.[44] Also, the researcher reviewed classified Operational Readiness Assessments (ORA) reports from two Iraqi Brigades located in Ramadi and Habbaniyah. The percentage of on-hand personnel to the MTOE was 110 and 118 percent, respectively.[45] Furthermore, based on the data, it is important to mention that the entire IA grew in personnel strength over the course of the two years. In January 2006, the army had 106,900 personnel. By December 2007, the army had 194,233 personnel on the rolls.[46] (See Appendix C: Growth of the Iraqi Army). Along with personnel growth, the IA's pay administration improved over the two-year period as well.

Personnel Strength In Comparison to MTOE (2006–2007)

	% of Assigned/ MTOE	% of Present for Duty/MTOE	% of Fired per month/ MTOE
2006	65%	24%	4%
2007	114%	65%	2%

Figure 13

Pay Administration

Pay: 2006

Pay is a motivator for the Iraqi soldier.[47] Young Iraqi males join the IA so they can earn money to support their families. In 2006, Iraqi soldiers earned approximately 477,000 dinar (approximately $320 USD) per month.[48] Every month, the soldiers were authorized 10 days of leave in order for them to bring the money home to provide consistent financial support for their families. The pay, however, was inadequate to support the families.[49] Poor security conditions and the dire quality of life caused soldiers to begin to complain about the low pay. Evidence of this observation was found in an 11 June 2006 article in the *Stars and Stripes,* which stated, "Iraqis complained more about pay and food than they did about combat casualties."[50]

The complete pay cycle was complex and arduous as the following field description of the researchers' attempts to obtain pay for the IA employees illustrates:

> Because the process to receive payment required many levels of approvals with signatures and rubber stamps, the researcher had to hand-walk the rosters and paychecks to various offices within MOD to meet the approval process. To expedite the pay process, the researcher normally had the "ticket-in" within the MOD building and was able to bypass Iraqi security checkpoints within the MOD hallways and offices. These checkpoints were normally manned by MOD employees and they only allowed MOD civilian workers, senior Iraqi Army officers, and all U.S. military personnel to bypass through all office and hallway checkpoints. However, the guards did not allow the Iraqi pay representatives through the checkpoints because the pay representatives were never considered essential personnel. This caused confrontations between the pay representatives and the guards, as the pay representatives were always on a short time frame to complete the pay process. The guards, however,

were never concerned about the pay representatives' short time frame, which is why the situation became confrontaticnal. Once the researcher defused the tension of the situation, the pay representatives were able to bypass the checkpoints to complete the check approval process.

The final step in this part of the pay process was to obtain the Minister's signature. The researcher and the pay representatives would hope that the Minister was in the office that day to sign the check. On one occasion, the researcher was standing outside the Defense Minister's office with the check so the minister could sign it. (The irony of this situation caused the researcher to make the analogy of a field grade officer standing outside then-Secretary of Defense, Donald Rumsfeld's office, hoping he would come out to sign a Marine Regiment's payroll roster.)

Once the check was signed, the next task was to coordinate a coalition convoy to the bank to cash the check.[51] The researcher would have to coordinate a convoy supported from Joint Headquarters, MNSTC-I. Convoys to support bank runs were not a capability offered by MNSTC-I, so maintaining contacts and close personal relations (affectionately called "hook-ups" by military personnel) with MNSTC-I personnel was essential. Therefore, the researcher was able to acquire convoy support. In addition, the researcher drove an up-armored SUV to transport the cash and a coalition convoy supported the movement. Upon arrival to the bank, the convoy personnel, normally U.S. Army, provided security around the bank perimeter while the researcher and the Iraqi pay representatives conducted business in the bank. On two occasions, the bank did not have money in the vault. The security situation was so poor during this period that no money was transported from the Central Bank of Baghdad to the local bank. Therefore, the convoy returned back to the International Zone (IZ),

66   *Applications in Operational Culture*

and the operation would fall to the next day.[52]

When the check was finally cashed, six to eight potato sacks full of cash were loaded into the researcher's SUV, and the convoy would travel back to the IZ under heavy security. From there, the researcher and Iraqi pay representatives flew from Baghdad to Al Asad. On the following day, the MiTTs then convoyed the money to each battalion and the battalions had approximately seven days to fully pay the soldiers. After seven days, the pay representatives would then return the money and rosters to the BDE Headquarters where reconciliation of the books took place. Next, the researcher oversaw which soldiers did not get paid, soldiers who needed to be removed from the pay roster, and soldiers who were not getting paid according to rank. A complete roster and report (in both Arabic and English) were sent directly to MOD for updating, which needed to be completed by the 20th of each month. MOD was required to update the database, and the changes were supposed to be reflected on the following month's pay roster. However, there were many months that the MOD never updated the pay rosters, and soldiers continued to serve without getting paid.

Ultimately, the researcher spent approximately 70 percent of his tour of duty addressing BDE pay issues. The researcher also spent 25 percent of his time in Baghdad (MOD), fixing pay problems within the BDE.[53] Something or someone was needed to implement a better pay administration process. Ultimately, the researcher and other advisors took ownership of the pay process and, similar to Marine Corps leadership, MiTTs persevered to ensure every soldier in the Brigade was paid. The first task the advisors accomplished was to review the Brigade's payroll roster. In January 2006, the advisors found there were over 400 "ghost soldiers" that needed to be taken off the payrolls because the extra money was an avenue for skimming off the top among senior officers. (Ghost soldiers are soldiers officially listed in the IA pay system, but who are unaccounted for because of being absent without leave.) In addition, the advisors determined there were over 150 soldiers who were not getting paid, and some who had not been paid in over six months. This problem, along with the complete pay cycle, needed attention at all levels of the process, particularly within the Ministry of Defense.[54]

Pay: 2007

Between May and July 2006, the BDE began utilizing a local bank in al-Anbar to cash the BDE's payroll check. Utilizing the local bank in Hadithah allowed the BDE to take a more active role in its own payroll process with minimal support from Coalition forces. Unfortunately, security within Bagdad worsened during the summer of 2006, and the Minister of Finance declined to move large amounts of cash across the country. Therefore, the BDE took a step backward having to again use a bank in Baghdad.

In 2007, improved security in Baghdad allowed ministry officials the time and effort to focus on payroll administration, which resulted in significant improvements, and employees were able to go to work unafraid.[55] One example of improvement was that 98.5 percent of the soldiers were paid and the "ghost soldier" count remained at a mere 40 per month by the end of November 2007.[56]

Since security improved throughout the country, the Iraqi Government's confidence grew to allow monetary funds to travel safely throughout the country. Therefore, MOD reevaluated the possibility of using local banks in Fallujah, Ramadi, and Hit to support the IA. In addition, MOD employees began using a direct deposit system. The system was experimental within MOD; if it was proven valid, direct deposit would then be implemented to the IA. Therefore, an Iraqi soldier could withdraw his money from his local community bank instead of relying on an antiquated pay system with multiple layers of procedures.[57] Furthermore, The Independent Commission on the Security Forces of Iraq reported:

> The MOD also continues to develop a new banking facility that is already providing roughly 2,800 employees with direct deposit services. The MOD is working to add another 1,500 personnel to the bank's rolls. The new MOD bank will facilitate payment while also reducing the risks of corrup-

tion in the payroll system. Although at present only a fraction of MOD personnel are part of the new direct deposit system, it is an important step in the right direction.[58]

Iraqi Brigade (1st BDE, 2d Div) Progress in Mosul, Iraq

The experiences of 2/7BDE were not atypical. The researcher was able to corroborate further evidence and gained additional information from an interview with a U.S. Army officer, Captain George T. Jones, who served on another MiTT and recently returned from Iraq. Captain Jones served as a Maneuver Trainer for 1st Brigade, 2d Iraqi Army Division, located in Mosul, from February 2007 until January 2008. During the interview, which was held on 13 January 2008, the researcher was able to establish a pattern consistent with the observations of the researcher for 2/7BDE.[59]

With regard to the three main areas discussed in this paper (i.e., logistics, pay, personnel), Jones stated that the surge had provided the space and time for the IA to operationally develop by "getting into the fight."[60] He stated that the IA participated in the surge alongside Coalition forces and determined that the IA was able to improve its performance because the soldiers were constantly challenged tactically and operationally by Coalition forces.[61] With regard to logistics, Captain Jones stated the BDE's life support was contracted by a local supplier outside of Mosul. Jones said that the BDE was being fed daily and that food supplies never fell short; however, Jones also stated that supply, maintenance, and medical requisitions remained a challenge because of MOD's systemic problems and its inability to support the Army. Therefore, Coalition forces still provided supply, maintenance, and medical support to the BDE.[62] This is consistent with the researcher's and Major Sims' reports.

On the other hand, the BDE's personnel strength remained steady during Jones' tenure. Recruitment of Iraqi males was successful, but MOD remained slow including/entering the new soldiers into the payroll system. According to Jones, "ghost soldiers" were still a problem during the course of the year, but his team was confident that the new Iraqi data base, Human Resource Information Management System (HRIMS)[63] was going to solve a majority of those issues. Jones also stated that the HRIMS would put a check and balance into the payroll system and identify whether the system was circumvented. In addition, Jones and his team dealt with corruption within the senior ranks of the BDE during the course of the year. However, his team was confident that the BDE could independently operate and sustain itself the following year, regardless of the Iraqi culture and mindset.[64]

Discussion

This data reflects and supports the argument that the surge and the uprising of Iraqi citizens (al-Anbar Awakening and the Concerned Local Citizens) allowed the space and time for MOD to develop its programs and policies to build, operate, and sustain the IA. The surge and the uprising of Iraqi citizens also gave Coalition forces the additional breathing room needed to enhance the IA's capability to independently sustain itself. Evidence collected from four different sources showed a pattern of consistency regarding the ways that the IA improved in the areas of logistics, personnel, and pay administration over the course of 2007. Although the speed of improvement was not a factor the researcher intended to study, it is important to mention that the data indicated that progress continued faster at tactical level units than it did at the strategic or national level.

Operational Readiness Assessment (ORA) reports from four BDE's were reviewed in December 2007 in the areas of logistics, personnel, and pay. Each MiTT Chief rated the Iraqi unit with high ratings in the areas of logistics, personnel, and pay.[65] All subjective comments regarding the three areas had a consistent message that MOD needed to accelerate its programs to support the IA, especially in administration and logistics.[66] The ratings from the four ORA for the four different BDE's indicated a consistent pattern that progress had been made at the tactical level and that MOD lagged at the strategic level. At the current pace, the lack of programs initiated and implemented by MOD will only stall Iraqi units from achieving ORA Level I,[67] (i.e., the ability for an Iraqi unit to plan, execute, and sustain a counterinsurgency). Furthermore, Sims indicated that MOD remained inefficient and continued its social bureaucracy and failure to plan."[68] Two critical shortfalls that have resulted in the IA's lack of strategic development and that have prevented the IA from fully operating independent of Coalition support are air capability and leadership development.

Air Capability

The IA's air lift capability is still limited. In 2005, Coalition forces completely eradicated the Iraqi Air Force and there were no flyable aircraft within the Iraqi Army's inventory. Coalition forces had to start from scratch to build a new Iraqi Air Force. U.S. Air Force Brigadier General Robert R. Allardice, in a report to the Foreign Relations Council, projected that the Iraqi Air Force will be fully mission-capable by 2012.[69] This projection has a direct impact on the overall movement and logistical capability the IA needs to operate independently.[70]

Leadership Development

The IA's officer and enlisted leadership development continues to progress, albeit slowly. However, it needs to be recognized that developing junior officers and enlisted noncommissioned officers (NCO) takes time. When Coalition forces disbanded Saddam Hussein's military, the foundational leadership within the former regime was also disbanded. Coalition forces then had to start from the bottom up to build a new officer and enlisted corps. It was not until late 2006 that Coalition forces allowed former officers and enlisted back into the IA. The resident knowledge of the former officers will continue to serve purposefully in developing the IA, but a new leadership culture will take many years to fully mature. Solid Iraqi leadership at senior ranks and elimination of government and ministry corruption, coupled with professional military education will be the cornerstone for a legitimate and credible IA.[71]

Recommendations

Given the evidence presented, the following are the researcher's recommendations on what Coalition forces and the Iraqi Government must focus on and accomplish in order for the Iraqi Army to become self sufficient and effective. These recommendations are not limited to U.S. military efforts, but include necessary actions and coordination with U.S. interagency organizations.

Logistics

Logistics development will continue to be the Achilles heel for the IA, especially at the strategic level. However, logistics improvement contributed to the Iraqi Army's ability to receive an ORA Level 2 performance rating by Coalition forces in 2007, even though the IA had received poor performance ratings the previous year. Coalition forces and MOD remain confident that if the strategic level of logistics progresses, the entire logistics system will become integrated and supportable to the Iraqi Army's operational and tactical units. In addition, the medical capability, which is the largest constraint for the Iraqi Army, continues to remain a long challenge for coalition forces. Therefore, Coalition advisors must continue to mentor and train senior MOD and Iraqi Army officers in strategic logistics to support the operational units, and must develop a synchronized logistical program that sustains the entire Iraqi Army

Personnel

The Ministry of Defense is meeting its desired manpower levels. It is unreasonable, however, to predict that MOD should curb its policy regarding the IA's leave process. Even though the U.S. feels that the leave policy is excessive (i.e., 10 days off for every 20 days of duty), the IA's morale is based largely upon the vacation time the soldiers receive by serving in their army. Implementing a western culture's style of leave system whereby one earns leave for time served would be detrimental to the Iraqi Army at this time. In addition, MOD must continue to build its legitimacy in order for personnel strength to continue to grow. Soldiers and Iraqi citizens must have confidence in their government and U.S. Interagency personnel and advisors must ensure that MOD establishes legitimacy to build that confidence.

Pay

To date, there are high expectations that the new HRIMS will significantly benefit the IA. The Ministry of Defense should continue to use local banks and a direct deposit program for the army. This process eliminates the burden of Coalition forces having oversight of the payroll process and puts the burden of control directly into the hands of MOD and the IA. If the IA is independent of Coalition forces' administering payroll, it will allow the advisors to focus on other critical areas of the IA, such as expanding their counterinsurgency abilities

Operationalizing the IA

Iraqi units in al-Anbar are currently assessed at ORA 2. At that level, it is important the MiTTs begin advising Iraqi units to focus more on the counterinsurgency fight. In the future, the strategy for the MiTTs should be the integration of the IA into fighting a counterinsurgency and to achieve operational independence while obtaining battlespace. According to an AAR by 3-1 BDE team chief:

> MiTTs must leverage the Iraqis across three levels of Logical Lines of Operations (LLOs): security, governance, and reconstruction. IA (Iraqi Army) officers are good at communicating with local nationals to find out real needs of the area. Get them involved. Too many IA units are simply relegated to patrolling in zone and following behind CF (Coalition Force) formations.[72]

Operational plans for the IA should focus on the areas of tactical and operational intelligence, fires/effects and information operations, humanitarian assistance, civil-military operations, and close air and artillery support.

Along with operationalizing the IA, it is important that the Iraqi Army and its officers begin solving their own problems. Sims, the researcher's successor, stated that the Iraqis should not be "baby sat."[73] Furthermore, advisors do not need to model the IA according to western military standards. Approaching the Iraqis with a western mind-set has caused frustration among advisors in the last two years.[74] Brigadier General David G. Reist is quoted in an article in the *Marine Corps Gazette* that "the tribal culture is the most complex issue in al-Anbar. The complexity of this issue stems from the simple fact that the Western mind does not (and may never) understand it. We do not have to understand though; simply accept it for what it is."[75]

It is logical to conclude that the IA has a distinct cultural advantage when it relates to tribal engagements. As security conditions improve, the U.S. military should start putting the Iraqis in front and let them figure out the problems themselves. On 10 January 2008, the *Washington Post* reported, "After countless unsuccessful efforts to push Iraqis toward various political, economic and security goals, they [U.S. military and diplomatic officials] have decided to let the Iraqis figure things out themselves." [76]

Iraqi Police

While initial decisions to disband the Iraqi Army were flawed, the decision to disband the Iraqi Police also had a wide impact on the overall security within Iraq.[77] As Coalition forces transitioned to a counterinsurgency, there were no local constables to provide security, civil control, or law and order. Therefore, when the regime fell, chaos ensued. Although the Iraqi Police was not considered a variable at the beginning of this study, it is important to mention that the police development has lagged behind the IA in their life support and pay administration. Like the IA, the quality of life of the Iraqi Police is also very important. Poor pay and life support leveraged the insurgency to influence young males. Therefore, Coalition forces and the Interagencies must ensure that police protection and security remains a high priority and that their life support programs and training requirements are met.[78]

Quality Advisors

Recently, U.S. Army General Barry R. McCaffrey (Ret) conducted an assessment of the situation in Iraq for the United States Military Academy. In his report, General McCaffrey was optimistic about the outcome of the status of the Iraqi Security Forces. He stated,

"The embedded U.S. training teams have simply incredible levels of trust and mutual cooperation with their Iraqi counterparts. ... This is the center-of-gravity of the war."[79] The researcher supports General McCaffrey's argument. Nevertheless, home-base units must select only the best qualified leaders for advisor billets. If the advisors are the center-of-gravity for winning this war, then units must not fall short in their personnel selection.

Another important factor is the size of the advisor teams. Coalition forces realized that the size of the teams had to increase in order to be effective for the IA. For example, "In al-Anbar alone, there were 1,700 advisors committed to the Iraqi Army, police, and border forces. This represented a 40 percent increase in advisor support."[80] Also, home station units must ensure that advisors continue with quality pre-deployment training and cultural development. Team Chiefs must hold strong leadership skills and must be able to develop the relationships not only between the IA, but also among partnered coalition units. Therefore, it is vitally important that the selection criteria be set at a high standard when selecting team members.

Conclusion

If the U.S. employed the same surge strategy in 2003 that was employed in 2007, conditions in Iraq may have been different. Specifically, the country may have seen a more stable government, and the IA and security forces may have been more operationally capable, Coalition forces and the IA may have sustained fewer casualties, and the U.S. troop strength may have reduced over time, thus allowing for shorter deployments. However, after four years of occupation, it was not until late December 2006 that the United States realized more forces were needed to meet the clear, hold, and build strategy.

The 2007 surge not only decreased the level of violence, but it was the catalyst that the Iraqi Army needed in order to improve in the areas of logistics, personnel, and pay administration. As this chapter has demonstrated, the Iraqi Army has made progress in these critical areas, which contributed to building the confidence and performance of the Iraqi Army in 2007. As the difficult work for advisors continues, major areas within the Iraqi Army, such as intelligence, air support, and medical capability, still need to be developed.

Successful development of the Iraqi Army in 2007 has been the starting point for the advisors' challenges in the coming years.

Troop levels may reduce, but the advisors' mission remains critically important. The U.S. must maintain close supervision and evaluation of the Iraqi Army's performance. With decreasing U.S./Coalition troop levels it is imperative that the Iraqi Army obtain the freedom to operate and sustain operations, not necessarily by Western standards, but by Iraqi standards. Stability in Iraq will only continue if Coalition forces put the Iraqis in front of the fight. The Iraqi Army may not meet the criteria of ORA 1 by Western standards, but at least the Iraqis are taking the lead.

Finally, the lessons learned from advising foreign militaries sets the tone for a new paradigm shift for the U.S. military. As the U.S. military continues to be involved in global conflicts, a key mission will be advising disorganized militaries often ridden with corruption. This paradigm shift will force the U.S. military to restructure its organization to conduct the full gamut of nation building operations. Therefore, the lessons the United States learns from building the Iraqi Army will be critical to the success of future nation building efforts.

Notes

[1] Paul S. Frederiksen, Emily J. Fall, and Patrick B. Baetjer Emily J. Fall and Paul S. Frederiksen, "Iraqi Security and Military Force Development: A Chronology" (Center for Strategic and International Studies, 26 May 2006), 2 (http://csis.org/files/media/csis/pubs/060526_isf_chron.pdf).

[2] Ellen R. Girden, *Evaluating Research Articles from Start to Finish*, 2d ed. (Thousand Oaks, CA: Sage, 2001), 23.

[3] Norman K. Denzin and Yvonna S. Lincoln, *Handbook of Qualitative Research*, 3d ed. (Thousand Oaks, CA: Sage, 2000), 444.

[4] David C. Gompert, "U.S. Should Take Advantage of Improved Security in Iraq to Withdraw," *San Francisco Chronicle*, 2 December 2007 (http://www.rand.org/commentary/2007/12/02/SFC.html).

[5] James J. Wirtz, "The Exquisite Problem of Victory: Measuring Success in Unconventional Operations," in Joseph R. Cerami and Jay W. Boggs, eds., *The Interagency and Counterinsurgency Warfare: Stability, Security, Transition, and Reconstruction Roles* (Carlisle, PA: Strategic Studies Institute, December 2007), 276.

[6] Anthony H. Cordesman and Patrick Baetjer, *Iraqi Security Forces: A Strategy for Success* (Westport, CT: Praeger Security International, 2006), 58.

[7] Ibid., 51.

[8] Ibid., 202.

76 *Applications in Operational Culture*

[9] Ibid., 202.

[10] U.S. House of Representatives, Committee on Foreign Affairs and the Committee on Armed Services. Situation in Iraq, General David H. Petraeus, Commander, Multi National Forces-Iraq, 10-11September 2007 (hereafter Petraeus testimony, 10-11 September 2007).

[11] Ibid.

[12] Andrew Tilghman, "Iraqis in Al Anbar Province Leaving Army in Droves," *Stars and Stripes* (Mideast edition), 11 June 2006 (http://www.stripes.com/article.asp?section=104&article=36968&archive=true).

[13] Petraeus testimony, 10-11 September 2007.

[14] Petraeus testimony, 10-11 September 2007.

[15] Kimberly Kagan, "The Al Anbar Awakening: Displacing al Qaeda from Its Stronghold in Western Iraq," WeeklyStandard.com, 2007 (http://www.weeklystandard.com/weblogs/TWSFP/IraqReport03.1.pdf).

[16] "Officials: Iraq Outlook Optimistic," *Time*, 26 December 2007.

[17] David Ignatius, "Paradox for Petraeus," Washington Post, 21 December 2007.

[18] Walter Pincus, "Wars Cost $15 Billion a Month, GOP Senator Says," *Washington Post*, 27 December 2007, A7.

[19] Joshua Partlow, "Iraq Safer but Still Perilous At Year-End, Petraeus Says," *Washington Post*, 30 December 2007, A23.

[20] Gompert, "U.S. Should Take Advantage of Improved Security in Iraq to Withdraw."

[21] Author interview with Maj Steve Sims, 11 December 2007.

[22] Ibid.

[23] Headquarters U.S. Marine Corps, *Warfighting*, MCDP 4 (Washington D.C: U.S. Marine Corps, February 21, 1997), 6.

[24] U.S. Government Accountability Office, *Operation Iraqi Freedom: DOD Assessment of Iraqi Security Forces' Units as Independent Not Clear Because ISF Support Capabilities Are Not Fully Developed* (Washington D.C: U.S. Government Accountability Office, 2007), 6.

[25] Ibid.

[26] Gen James L. Jones (USMC) et al., *The Report of the Independent Commission on the Security Forces of Iraq* (Washington, DC: Center for Strategic and International Studies, 2007), 54 (online at http://csis.org/files/media/csis/pubs/isf.pdf).

[27] Ibid., 53.

[28] Tilghman, "Iraqis in Al Anbar Province Leaving Army in Droves."

[29] U.S. Department of Defense, *Measuring Stability and Security in Iraq in Accordance with the Department of Defense Appropriations Act 2007, Section 9010, Public Law 109-289* (Washington, DC: U.S. Department of Defense, December 2007), 14.

[30] Col Thomas C. Greenwood (USMC), interview with author, 14 January 2008.

Developing the Iraqi Army   77

³¹ Jones, *Report of the Independent Commission on the Security Forces of Iraq*, 53.

³² Sims interview.

³³ Ibid.

³⁴ Tilghman, "Iraqis in Al Anbar Province Leaving Army in Droves."

³⁵ Ibid.

³⁶ Carter Malkasian, "Thin Blue Line in the Sand," *Democracy Journal*.Org (Summer 2007), 30.

³⁷ Maj John Bilas (USMC), review of classified operational readiness reports from MNF-W SIPR Web site, 31 March 2006 (hereafter Bilas, [date]).

³⁸ Bilas, 1 April 2006.

³⁹ LCpl Joseph D. Day and Spc Ricardo Branch, "Anbar Leaders Celebrate Awakening," *Marine Corps News*, 8 July 2007.

⁴⁰ Ibid.

⁴¹ Bilas, 12 February 2006. The Iraqis were authorized 10 days of paid leave every 20days. As a result, approximately 30 percent of the BDE was on authorized leave at a time. MOD contracted passenger buses to transport soldiers from Al Asad to Baghdad. Coalition forces provided security for the buses. In 2007, the Iraqi BDE started to provide their own security, thereby eliminating the security requirement for Coalition forces

⁴² Sims interview.

⁴³ Ibid.

⁴⁴ Ibid.

⁴⁵ Bilas, 21 December 2007.

⁴⁶ Size of Iraqi Security Forces on Duty (Table), "Iraq Index, Tracking Variables of Reconstruction and Security in Post Saddam Iraq" (Brookings Institution, Washington, DC, 21 December 2007), 31 (http://www.brookings.edu/saban/~/media/Files/Centers/Saban/Iraq%20Index/index20071221.pdf).

⁴⁷ Center for Advanced Operational Culture Learning, "Final Report, 3d Battalion, 2d Brigade, 7th Division MiTT," 6 December 2006, 35.

⁴⁸ Bilas, 17 April 2006.

⁴⁹ Bilas, 12 February 2006.

⁵⁰ Tilghman, "Iraqis in Al Anbar Province Leaving Army in Droves."

⁵¹ Bilas, 17 April 2006.

⁵² Ibid.

⁵³ Ibid., 16 December 2006.

⁵⁴ Ibid., 12 February 2006.

⁵⁵ Sims interview.

⁵⁶ Ibid.

⁵⁷ Ibid.

⁵⁸ Jones, *Report of the Independent Commission on the Security Forces of Iraq*, 51.

78 Applications in Operational Culture

[59] Capt George T. Jones (USA), telephone conversation with author, 13 February 2008.

[60] Ibid.

[61] Ibid.

[62] Ibid.

[63] MOD established a new database called the Human Resource Information Management System (HRIMS). This data base system allowed changes to pay rosters at the BDE level. The BDE G-1 could input changes to the system in al-Anbar and then transmit that data change directly to MOD in Baghdad. This information management system made the reconciliation process faster and soldiers were paid in a timely manner.

[64] Jones telephone conversation, 13 February 2008.

[65] Bilas, 21 December 2007.

[66] Ibid.

[67] Ibid.

[68] Sims interview.

[69] BGen Robert R. Allardice (USAF), podcast interview by Greg Bruno, Council on Foreign Relations, 5 February 2008 (http://www.cfr.org/publication/15421/).

[70] Ibid.

[71] Simms interview.

[72] MiTT 1/1 AAR, 11 September 2007.

[73] Sims interview.

[74] Ibid.

[75] BGen David G. Reist (USMC), "Twelve Things I Wish I Had Known," *Marine Corps Gazette*, October 2007, 77.

[76] Thomas E. Ricks and Karen DeYoung, "For U.S., The Goal Is Now 'Iraqi Solutions,'" *Washington Post*, 10 January 2008, A1.

[77] Cordesman and Baetjer, *Iraqi Security Forces*, 10.

[78] U.S. House of Representatives, Committee on Armed Services, "The Continuing Challenge of Building the Iraqi Security Forces," 20 June 2007, 6.

[79] General Barry R. McCaffrey (USA Ret), "After Action Report, Visit Iraq and Kuwait 5-11 December 2007" (United States Military Academy, 18 December 2007), 5 (online at http://www.afa.org/EdOp/edop_12-27-07.pdf).

[80] Zaid Sabah and Ann Scott Tyson, "Security Pact on Iraq Would Set U.S. Exit," *Washington Post*, 11 December 2007, A15.

Chapter 4

The Way Ahead: Reclaiming the Pashtun Tribes through Joint Tribal Engagement

Major Randall S. Hoffman, USMC

Major Randall S. Hoffman is a Marine infantry officer with more than 24 years of service. He has served from private to major in various infantry, reconnaissance, and special operational units and held several leadership and command positions throughout his career. He graduated with honors from Indiana University in 1994 with a bachelor of arts in history. He is a recent distinguished graduate of the Marine Corps Command and Staff College, attaining a masters in military studies. He is currently a student in the Marine Corps School of Advanced Warfighting.

From 2003 to 2005, Major Hoffman served as a senior military advisor in both Operations Enduring Freedom and Iraqi Freedom. His operational experience includes tours in the Persian Gulf, Liberia, the Congo, Sierra Leone, Iraq, and Afghanistan. In 2005, Major Hoffman was assigned as the inspector-instructor and company commander of Company K, 3d Battalion, 24th Marines, in Terre Haute, Indiana.

The United States' enemies, al-Qaeda and the Taliban, freely operate in a sanctuary within the North-West Frontier Province (NWFP) and the Federally Administered Tribal Areas (FATA) of Pakistan where they launch attacks into Afghanistan at a time and place of their choosing. The FATA and NWFP combined comprise 101,742 square kilometers of some of the most treacherous terrain in the world and are home to over 22 million people, predominantly of Pashtun origin. Historically, these regions have served as a training ground, a logistics base, and a conduit for supplying Mujahideen fighters against the former Soviet Union. Since 2001, al-Qaeda and the Taliban have expanded their operations against the U.S. and NATO forces stationed in Afghanistan by utilizing key

80 *Applications in Operational Culture*

partnerships with the Pashtun tribes living in the NWFP and FATA.

Pakistan has never held governance over the NWFP and FATA due to the Pashtun tribal reluctance to accept outside rulers who possess cultural and social structures that are different from their own. Added to this xenophobia is that a very large number of the Pakistani military are of Pashtun decent and unwilling to wage war against their own tribe.

Afghanistan, the NWFP, and the FATA are complex regions with cultural, economic, political, and religious landmines at every turn. Nevertheless, through careful study of the region's history, social structures, and a solid partnering effort amongst the U.S. military, Coalition forces, the Interagency, Pakistani military, and the Pashtun tribal leadership, the United States can achieve its goals of destroying al-Qaeda and the Taliban without destabilizing the region.

Pashtuns: The Largest Ethnic Group in Afghanistan

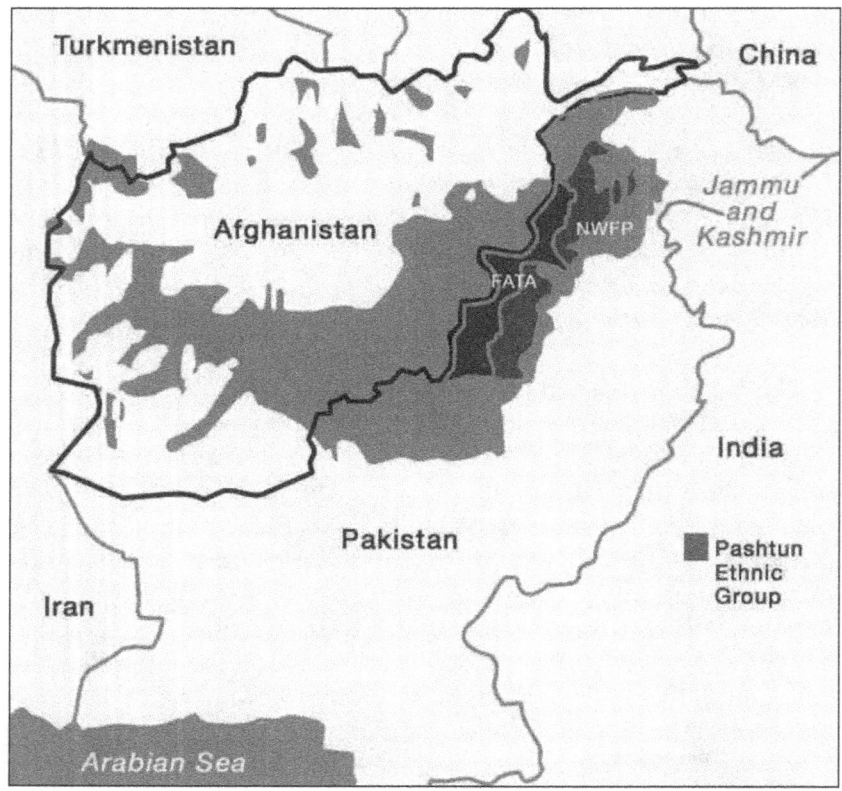

Figure 14

Figure 15
Pakistan's Federally Administered Tribal Areas

The United States must begin strategic and operational planning for the FATA and NWFP regions, as well as eastern Afghanistan, to defeat al-Qaeda and the Taliban in order to bring security and stability to the region.

This chapter argues that the Pashtun tribes residing within eastern Afghanistan, the FATA and NWFP, are the enemy's true center of gravity (CoG). In order to effectively attack this center of gravity, the author proposes a new strategy that utilizes all the U.S. elements of power—diplomatic, information, military, economic—integrated with an operational culture understanding (DIME-OC) as a ways and means to destroy al-Qaeda's sanctuaries while diminishing the relevance of the Taliban among the Pashtun people. The author's proposed strategy is based on personal interviews with Pashtun tribal leaders, and two years of experience gained serving alongside Pashtuns as a military advisor from 2003 to 2005. During his deployment the author employed, with significant success, some of the operational culture methods described in the following pages.

Afghanistan at the Crossroads

When we entered Afghanistan in 1979, people gave us a very nice welcome. Exactly a year later, 40 percent of the population began to hate us. Five years later, 60 percent of the population hated us. And by the time we were to pull out, 90 percent hated us. So we understood, finally, that we are fighting the people.

-LtGen Ruslan S. Aushev, USSR,
after serving five years in Afghanistan[1]

The Taliban are well aware that the center of gravity in Afghanistan is the rural Pashtun district and village, and that the Afghan army and coalition forces are seldom seen there. With one hand, the Taliban threaten tribal elders who do not welcome them. With the other, they offer assistance.

-Thomas H. Johnson and M. Chris Mason,
Atlantic Monthly (October 2008)[2]

Although the United States has achieved some of its original objectives in Afghanistan, the Taliban and al-Qaeda continue to maximize almost every principle of war against the United States with greater lethality each year. The International Security Assistance Force (ISAF) Commander, General David D. McKiernan, witnessed a "historic level of militant sanctuaries in the tribal areas of Pakistan that have fueled the insurgency not only on the Afghan side of the border but the Pakistani side of the border."[3] U.S. strategy has failed to compel the majority of Pashtun tribes living within

Afghanistan, the Federally Administered Tribal Area (FATA) and the North Western Frontier Province (NWFP) of Pakistan to cease their support of the Taliban and al-Qaeda. The U.S. must understand that the Pashtun people are the enemy's true center of gravity and that without their support, the enemy will eventually collapse. This chapter will explain why the Pashtun tribes are the CoG, how the United States can engage the Pashtuns through the use of Joint Tribal Teams, and the importance of utilizing social, economic, political, and Islamic educational engagement within Afghanistan and Pakistan to meet Pashtun needs while winning the war against al-Qaeda and the Taliban.

Tribal Engagement: The Center of Gravity

When you're wounded and left on Afghanistan's plains, And the women come out to cut up what remains, Jest roll to your rifle and blow out your brains, An' go to your Gawd like a Soldier.

- Rudyard Kipling, "The Young British Soldier" (1892)

Joint Publication 5-0, *Joint Operation Planning*, defines the CoG as "a source of moral or physical strength, power, and resistance"—what Clausewitz called the "the hub of all power and movement, on which everything depends...the point at which all our energies should be directed."[4] In *Clausewitz's Center Of Gravity: Changing Our Warfighting Doctrine-Again!*,[5] Antulio Echevarria strips Clausewitz's CoG down to the bone in his deciphering of Clausewitz's true definition of the center of gravity, as "the one element within a combatant's entire structure or system that has the necessary 'centripetal' force to hold that structure together." Most important for U.S. operations in Central Asia, is Echevarria's conclusion that Clausewitz was not referring to a specific enemy strength or weakness but a "focal point." Echevarria suggests, "A blow at the enemy's CoG would throw him off balance or, put differently, cause his entire system (or structure) to collapse." Eschevarria concludes that in "Clausewitz's effects-based approach, the effect and the objective—total collapse of the enemy—were always the same." Echevarria points out the relevance of effects-based operations (EBO) for 21st century military planners, referring to General Anthony C. Zinni's comments "that they [EBO] were akin to dissolving 'the glue' that holds a table together, rather than striking at its individual legs."[6]

The Center of Gravity is the Pashtun

Today, the "glue" holding the Taliban and al-Qaeda together are the hundreds of Pashtun tribes (*tabars*) and kinship groups (*khels*) residing in the FATA and NWFP regions of Pakistan. This is the region where the enemy lives, recruits, trains, and reconstitutes itself so it can launch further attacks against the U.S. and coalition forces within Afghanistan. An unnecessary gamble and, potentially, insurmountable obstacle to success in the current U.S. strategy is its over-reliance on building a strong central Afghan government in the hopes that its growing military strength will become strong enough to strike the individual "legs" of the al-Qaeda-Taliban "table" when one or two of the "legs" cross into Afghanistan. Instead, U.S. policy and strategy should focus on "dissolving the glue" of Pashtun tribal support, through the building of tribal partnerships regardless of whether the al-Qaeda-Taliban table is resting within Pakistan (FATA-NWFP) or elsewhere.

The importance of gaining Pashtun tribal support in the FATA-NWFP can be seen from the lessons learned by the Soviets. The Soviet Army tried to fight the Pashtuns by dominating the majority of the lines of communications, cities, and villages in the lower lands of Afghanistan. The Soviets ruled out operating in the FATA and NWFP and most likely viewed operations in that region as impossible due to their reluctance to expand the war into Pakistan. As a result, their strategy of dominating urban areas and road networks within Afghanistan ultimately failed because they allowed a safe haven to exist within the FATA and NWFP. From there, the Mujahideen, predominately Pashtun tribesmen from the hill tribes (Ghilzai-Karlanri), trained, equipped, and launched attacks against the Soviets inside of Afghanistan. Once the Mujahideen returned safely across the Durand Line, they reconstituted their forces and prepared for more engagements.

The Soviet Army realized too late in the Soviet war in Afghanistan that "the main methods of fighting the armed opposition could not be military actions by regular forces." Instead, they adopted a ninth-inning strategy of "determined social-economic, political, and organizational-propagandistic methods."[7] Similarly, after eight years, the United States has not attained an acceptable level of security within Afghanistan. Even if the United States were to surge four divisions (108,800 personnel), as the Soviet 40th Army did in 1985 (resulting in their highest number of combat deaths in the war),[8] it would not address the enemy's true center of gravity.

Although supporting the growth of an Afghan government is vitally important to the future stability of the country, the U.S. strategy is not focused on the right level of governance. U.S. strategy must focus less on growing the Afghan central government from its top leaders (mostly Durrani Pashtun) on down, and focus more on tribal relationship-building and local community governance with the other, lesser known Pashtun tribes as well as other ethnicities, from the bottom up.

The recent U.S. Army and Marine Corps' manual *Counterinsurgency* (FM 3-24, MCWP 3-33.5) stated, "The primary objective of any counterinsurgency (COIN) operation is to foster development of effective governance by a legitimate government."[9] Currently, a growing number of Afghan ethnicities and tribes below the district level of government do not accept the Afghan central government as a legitimate government.[10] This is primarily because Afghanistan has once again followed its historical path of "person-centered, sovereignty-based rule over an emasculated body of mistrusting subjects."[11]

In any COIN effort, "every aspect of sound counterinsurgency strategy revolves around bolstering the government's legitimacy. When ordinary people lose their faith in their government, then they also lose faith in the foreigners who prop it up."[12] Today in Afghanistan, the FATA, and the NWFP, the vast majority of what possibly might be the "largest tribally organized group in the world,"[13] the Pashtuns, do not recognize the Afghan government as being legitimate. The United States must not make the same mistake that the Soviet Union did in supporting a government that was not legitimate in the eyes of most Afghans. Doing so only increases the likelihood that Americans may share the fate of Kipling's soldier on an Afghan plain.

The Pashtun: A House Divided

The Cultural Context of Pashtunistan

"The Pukhtun is never at peace—Except when he is at war"
Pukhtun (Pashtun) Proverb[14]

Afghanistan is made up of diverse ethnicities. The Pashtun have historically ruled the country; however, they are currently divided, both physically and politically. The most significant cultural and political division lies between the Afghan tribes that inhabit the rural hills and mountains of eastern Afghanistan and the FATA and NWFP of Pakistan versus the urban dwelling plains Pashtun. Divisions among the Pashtun are centered on the Ghilzai sub-tribe and the "Karlanri" (Pashtun hill-tribes)—which, in this article, refers to the warlike Afridis, Daurs, Jadrans, Ketrans, Mahsuds, Mohmands, and Waziris (North and South), along with many other smaller sub-tribes–and, their rivals, the Durrani Pashtuns, who occupy the southern, flatter portion of Afghanistan.

Pashtuns are a warrior class of people whose numbers are estimated to be between 20 to 24 million. Although anthropologists have long since discovered that Pashtuns are in fact a heterogeneous people, the Pashtuns hold firm to the belief that they descended from "one" founding father and therefore hold strong to the concept of a "one Pashtun people identity."[15] With that said, Pashtuns have serious divisions along the lines of social hierarchy, linguistics, economic systems, political rule, and religious beliefs that are a constant point of friction. These cultural divisions are most prominent between the Durrani Pashtun branch that inhabit the southern areas of Afghanistan bordering Baluchistan, and the Ghilzai and Karlanri Pashtuns who inhabit eastern Afghanistan in addition to the NWFP and the FATA of Pakistan.

Collectively, Pashtuns are one of the most segmented societies on earth and are broken down into approximately 350 tribes, which are divided further into hundreds of clans and thousands of sub-clans.[16] Pashtuns use tribal descent order to establish their "boundaries for social nearness and distance" and to mark the lines of "conflict and solidarity."[17] As a Pashtun related to a researcher, "If I see two men fighting, I am supposed to side with the one who is 'closer' to me," that is, "the one with whom I share the nearest common patrilineal ancestor."[18] Another way to examine Pashtun social-political relationships is to visualize M. Jamil Hanifi's graphical representation of the Afghan [Pashtun] individual as being:

> surrounded by concentric rings consisting of family, extended family, clan tribe, confederacy, and major

cultural-linguistic group[s]. The hierarchy of loyalties corresponds to these circles and becomes more intense as the circle gets smaller...seldom does an Afghan, regardless of cultural back-ground, need the services and/or facilities of the national government.[19]

M. Nazif Shahrani, chairman of Near Eastern Languages and Cultures and Middle Eastern and Central Asian Studies, Indiana University, examined the "erosion" of the outer limits of these circles and argues that Afghanistan's rough political history created a situation in which "consistent policies and practices of political mistrust directed against the great majority of Afghans [has] promoted a general attitude of distrust of politics and politicians." He goes on to write that these "experiences...have weakened traditional communities [i.e., civil society]," and has caused a "general erosion of the social capital of trust beyond the circles of family and close kinsmen or at most one's [Pashtun] own ethno-linguistic group." Future U.S. strategy, therefore, must be focused on the point where this trust was broken, at the inner circle, in order to begin its rebuilding from the only functioning Pashtun circle left, the tribe.[20]

The Pashtun: A House Divided

"We are content with discord, we are content with alarms, we are content with blood...we will never be content with a master."

- Elderly Pashtun tribesman talking to British official Mountstuart Elphinstone in the 1800s

The Durrani tribe has been the dominant Pashtun branch that has supplied Afghanistan with its ruling families since the Durrani Empire in 1747.[21] Durrani tribal leadership and social structures differ considerably from that of the Ghilzai and Karlanri with the Durranis, primarily large-scale agrarian cultivators, adhering to a more heirarchical view of tribal leadership. This may be traced to their pattern of private ownership of land instead of the communal sharing of land found among the Ghilzai and Karlanri Pashtuns. In contrast, the Ghilzai and Karlanri are nomadic and inhabit the hills and mountains of eastern Afghanistan, the NWFP, and the FATA, where arable land is scarce and economic livelihood difficult. As a result, both Ghilzai and Karlanri adhere to a stronger, more egalitarian form of tribal governance in which the tribal *jergah* rule is supreme and heavily influences the distribution of land for the benefit of the tribe and not the individual as well as many other important tribal decisions, such as war councils for blood feuds.

The greatest difference between the Ghilzai-Karlanri and the Durranis, however, is in the interpretation and degree of adherence to *Pakhtunwali* (The Way of the Pashtun), the Pashtun honor code.[22] *Pakhtunwali* dominates every aspect of the Ghilzai and Karlanri tribal life. Along with Islam, *Pakhtunwali* is the foundation upon which all Pashtun tribal affairs and social behaviors are built. Problematically, *Pakhtunwali* is often perceived by the United States as a code for housing one's enemies, which is an extremely narrow and short-sighted view when compared to the total effect that *Pakhtunwali* has in influencing every aspect of Pashtun tribal culture, especially among the rural, egalitarian Ghilzai and Karlanri in which:

> It is more than a system of customary laws, it is a way of life that stresses honor above all else, including the acquisition of money or property. It is a code that is practically impossible to fulfill in a class-structured society or in areas where "governments" prohibit such institutions as blood feuds and demand tax payments.[23]

The Ghilzai, as well as their Karlanri neighbors, understand themselves to be "the only true Pashtuns" because only they can maintain the strict standards of autonomy demanded by the *Pakhtunwali* code that drives the social structure of the people who:
> Inhabit the most marginal lands that are poor and beyond government control. A Pashtun family without honor becomes a pariah, unable to compete for advantageous marriages or economic opportunities, and shunned by other families as a disgrace to the clan. *Pashtunwali* also provides a legal framework for social interaction. [Pashtunwali] is still by all odds the strongest force in the

tribal area, and the hill [Pashtuns]...accepts no law but their own.[24]

After gaining a better understanding of how the hill tribes (Ghilzai and Karlanri) implement *Pakhtunwali*, one can understand the extreme difficulty of trying to dissuade several million of these fiercely loyal tribesmen to turn away or break faith with a social, legal, and political system that has defined them, in their folklore and as a people. For many of these tribes, *Pakhtunwali* has been the bedrock of their tribal ancestry for generations. *Pashtunwali* core tenants, which include "self-respect, independence, justice, hospitality, forgiveness, and tolerance"[25] cannot, and must not, be abandoned at any cost. This is not to imply that all Pashtuns embody this ideal, or adhere to it at the same level, but it is fair to say that when other Pashtuns see the Ghilzai tribesmen in lower, more urbanized regions of Afghanistan "with their long fighting knives visible in their waistbands, the townspeople are likely to sneak admiring glances and mutter something to their friends about 'real Pashtuns.'"

Consequently, the Ghilzai-Karlanri tribes do not accept any government law that supersedes their tribal law; a nonacceptance of outside rule that includes the Durrani-led, Kabul-based central government in Afghanistan. This Durrani-dominated rule has always required the tenuous support of those lesser, politically prominent Pashtun tribes, meaning those Durrani sub-tribes that were not in power or politically weak non-Durrani Pashtun tribes.

The Ghilzai-Durrani rift that al-Qaeda and the Taliban currently exploit with great effect is showcased in the recent history of the Taliban's military push into Mazr (illustrated in Carter Malkasian and Jerry Meyerle's *A Brief History of the War in Southern Afghanistan*). The Ghilzai, who supported the Taliban and Mullah Omar (a Hotak-Ghilzai Mullah) in their rise to power, were severely underrepresented on the Taliban's "military *Shura*" in which:

> Out of 17 members in the Kabul Shura in 1998, at least eight were Durranis while three are Ghilzai and only two were non-Pashtuns. After the [Taliban's] Mazr defeat in 1997 there was growing criticism from Ghilzai Pashtun commanders that they were not being consulted on military and political

issues, despite the fact that they now provided the "the bulk of military manpower.[26]

The lack of Ghilzai representation in the current Durrani government and the foreboding with which they view a future held in the hands of Durrani leadership suggests the accuracy of Malkasian and Meyerle's conclusion that:

> The Taliban regained influence in Zabul through their religious network and exploiting tribal rifts. Mullah Dadullah (Kakar tribe-Pashtun), supervised the Taliban's infiltration into Zabul. Events there foreshadowed the future of Helmand and Kandahar. Zabul was divided between the "majority" Ghilzai (Hotak, Khakar, and Tokhi tribes) and the "minority" Durrani Pashtuns. These Ghilzai tribes were inclined to oppose the [Karzai] government on the basis of the long-standing rift between the Ghilzai and Durranis. When Karzai replaced the governor in 2003, a Tokhi Ghilzai, with a Durrani, elements of that tribe stopped fighting for the government.[27]

Al-Qaeda and the Taliban today are growing the next generation of Pashtun fighters from among the thousands of young Ghilzai and Karlanri Pashtun boys. By building their Wahabi *madrassas* throughout the NWFP and the FATA, they are able to indoctrinate young Pashtun boys into Wahabi style Islamic fundamentalism whilst providing the necessary military training and equipment to wage war against the U.S. and coalition forces. This arrangement is the heart of the problem in Afghanistan today. The Taliban leaders within the FATA and the NWFP understand this fact better than do the U.S. strategists. They maintain their vital alliance with the Ghilzai-Karlanri tribes, investing immense effort into widening the gap between the Durrani and the rural hill tribesmen. The Taliban, with the assistance of al-Qaeda, exploit their geographical (safe-haven), kin-based, political and social advantage against the United States, Afghanistan, and their coalition partners by reaching out to these disenfranchised Pashtun tribes while amplifying their closeness both tribally and politically to the Ghilzai and Karlanri through patrilinial descent and social, [religious] and political nearness. By making this cultural-closeness connection and pointing out differences between themselves and the Durrani led rulers, the Taliban are able to solicit and attain the solidarity and allegiance of key tribal Pashtun leaders and with them, their fighting men, pushing the Pashtun base farther away from Afghanistan's goal of a unified country.

U.S. strategy in Afghanistan has failed to recognize the difficult reality of integrating various Afghan ethnicities and tribes, including the Pashtuns, into one Afghan governmental system. Despite the U.S. government's own challenges and failures in integrating Native Americans into a national government (ultimately requiring the establishment of the Bureau of Indian Affairs in 1824), few lessons have been learned. The U.S. government has never governed Native Americans peacefully without their own volition. Afghan tribes will be no different. Assisting the Afghan government in achieving legitimacy will be a far more difficult task than that of the Native American example due to extenuating factors in Afghanistan and Pakistan: multiple ethnicities, extreme poverty, lack of natural resources, history of civil war, ongoing tribal warfare, drug trade, outside influence on the country from other state and non-state actors, and the FATA and NWFP sanctuary issues within Pakistan.

Looking Backward For the Way Ahead: Britons and Pashtuns

Don't interfere with the tribal customs and, if you have to use force, use it quickly and effectively.

Major John Girling, British NWFP Veteran (2008)[28]

Studying the British Empire's struggle to secure the North Western Frontier NWF (now the NWFP) in the First Anglo-Afghan War of 1839-1842 provides important lessons on the role of the Pashtun. Great Britain went on to fight the Pashtun tribes twice more in the Second Anglo-Afghan War of 1878-1880 and in the Third Anglo-Afghan War of 1919-1921. In all three wars, the British learned valuable lessons that they recorded for students of military history and operational art especially in the realm of Pashtun tribal culture and the British's attempts to implement it into their military operations.

Nine years before Clausewitz completed *On War*, several lesser-known British military officers and political agents serving in the

92 *Applications in Operational Culture*

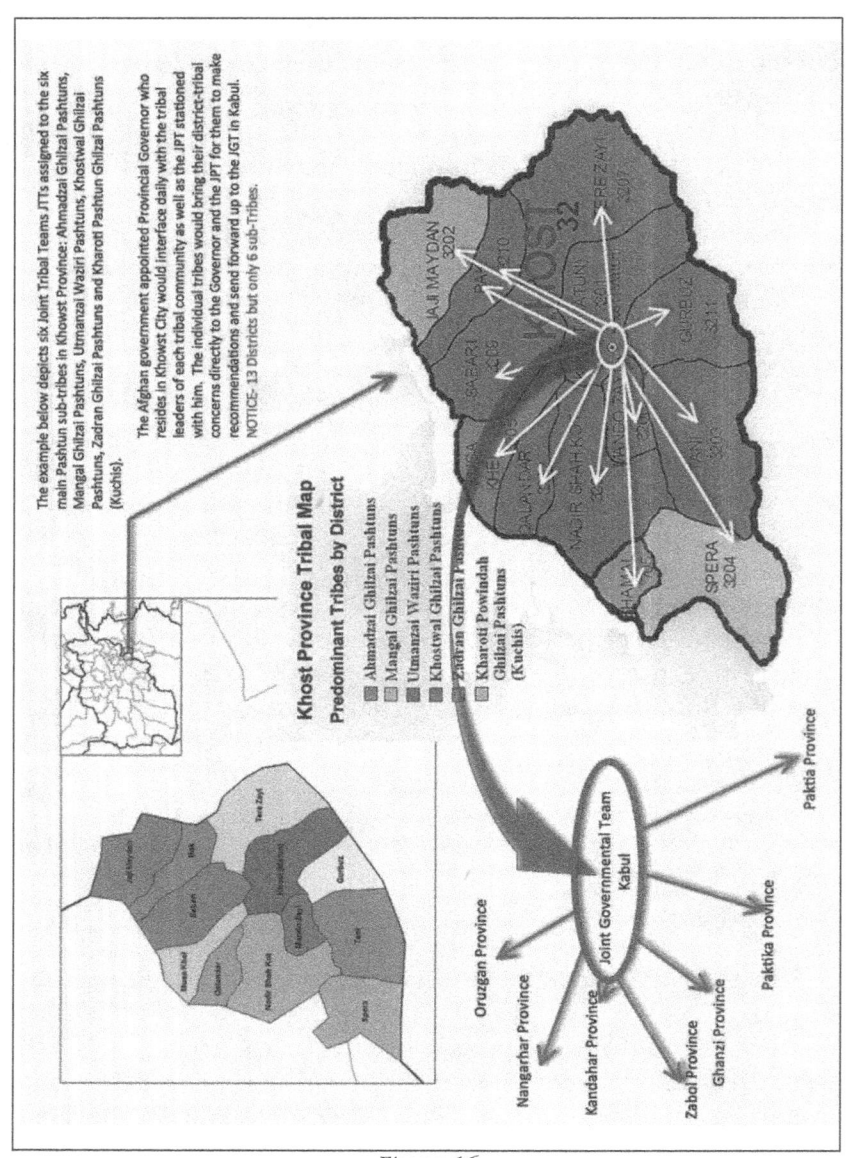

Figure 16
Khost Province Tribal Map

Indian-North West Frontier recorded their observations of the Pashtun tribes as they lived among them. Over a period of almost 150 years, these Britons documented their observations, all of which provide an "operational cultural" insight into the Pashtun.

The first of these men was Mountstuart Elphinstone, a British Indian official who in 1808 was placed in charge of a diplomatic mission to Kabul, Afghanistan, with the purpose of "encouraging the Afghans to resist the French."[29] His excellent reconnaissance of the Afghan tribes and rulers gave Great Britain the first look into the terrain, tribes, and cultures of those who inhabited the North West Frontier (See Figure 16). In his first account of Pashtun tribal governance, Elphinstone warned that since:

> Each [Pashtun] tribe has a government of its own and constitutes a complete commonwealth within itself, it may be well to examine the rise of the present situation of those commonwealths, before we proceed to consider them as composing one state, or one confederacy under a common sovereign.[30]

It is in Elphinstone's account of Pashtun governmental systems that the United States may see similarities to contemporary issues of governance in Pashtunistan. Elphinstone suggested caution whenever interfering with tribal governments that functioned in 1809 much as they do in 2009:

> The tribes continue in a great measure unmixed (each having its territory compact). They still retain the patriarchal government I have alluded to [Emir or King in Kabul (Durrani)], and the operation of the principle which I suppose to have separated them, is still very observable.[31]

James Atkinson, a British officer who served later (1839-1840) in Afghanistan, wrote of the Afghan Ruler, Dost Mohamed Khan (Durrani tribe, Barakzai kinship group) and his inability to subdue the rural Pashtun hill tribes under his government:

> His position as a usurper, or a conqueror, or whatever designation may be applied to seizing the government of Caubul [Kabul], an integral part only of the old empire, was surrounded with great difficulty; and the questionable nature of his authority,

added to the deficiency of his resources, compelling him to commit many acts of oppression, could not invest him with much substantial national power to resist any hostile encroachment from the west. He was even too weak to repress the turbulent and refractory spirit of the chiefs in remote districts, addicted as they were to plunder and rapine, so that their outrages became not only overlooked but sanctioned, on the reluctant and occasional payment of a precarious tribute, whilst the petty landholder was subject to every species of vexatious exaction[32]

Brigadier General Sir Percy M. Sykes, a soldier, diplomat, and scholar who served in the Indian Army and conducted campaigns in Afghanistan, wrote of the same problems Atkinson witnessed: the inability of Afghan rulers to not only govern but also to subdue the Pashtun rural tribes and their military strength. Sykes observed, 62 years after Atkinson, that:

> I have pointed out more than once that the real military strength of Afghanistan lies in its armed population rather than in its army, but the difficulties of supply strictly limit the numbers of armed tribesmen who can be kept in the field for any long period. Afghanistan, at this time, was divided into 10 military districts, all of them, except in the case of Kabul, being in touch with its frontiers. Its effectives were nominally 38,000 infantry, 8,000 cavalry and 4,000 artillerymen. They were badly trained, with obsolete guns and rifles, although their courage and endurance were beyond dispute.[34]

Lieutenant General Sir Sydney J. Cotton, K.C.B. deployed to India first in 1810 and later conducted operations in the North-West Frontier from 1854 to 1863. Considered "one of the best officers in India"[35] by Sir John M. Lawrence, British Governor-General of India, Cotton's account of attempting to govern the Pashtun tribal regions of the North West Frontier revealed an eternal Afghan truth in his conclusion that, "Civil government in such a country is, in truth, not only a stumbling block, but a manifest absurdity." Cotton, explains that through constant "intercourse with the "people" of the country, within, on, and beyond our border, he was able to glean intelligence to which few others were accessible. "Strange as it may appear, the civil authorities of India [British Command], are kept in the dark by the designing natives about them; and whilst military men, who possess no political power nor influence, can obtain intelligence, the civilians cannot."[35]

The British involvement in Afghanistan offers historical lessons for the United States. Irrespective of a strong ruling power in Kabul, whether monarch or president, the successful Afghan ruler will utilize some form of a tribally linked, monarchial governance to achieve his goals. Therefore, the United States should acknowledge that in Afghanistan, the tribe is the strongest form of a functioning government now. Through partnership with the various Pashtun tribal communities and their involvement in the political process, the United States may assist the Afghan government to achieve legitimacy amongst all Pashtuns. The ways and means to accomplish this is through the "existing frameworks of the tribal organization and tribal customs," as recommended by Lieutenant Colonel C.E. Bruce in *Waziristan 1936-1937: The Problems of the North-West Frontiers of India and Their Solutions*.[36] For Bruce, "The supreme test of any policy, if it is to be successful, must still be the welfare of the people—the welfare of the tribes—because any policy which has subordinated their welfare to purely political considerations has always failed." Bruce provided four timeless, essential tenets for governing Pashtunistan:

> 1. It must be built on existing frameworks of the tribal organization and tribal customs.
>
> 2. It must be worked through the headmen [leader of that community] because there is no efficient substitute and without him nothing can be done.
>
> 3. The cornerstone of the administration is the District Officer. The cornerstone of the tribal organization is the headman. Any policy that weakens their power weakens the very foundations of law and order, and causes lawlessness. And lawlessness means suffering to the law-abiding masses. It

96   *Applications in Operational Culture*

> therefore, fails to stand the supreme test—the welfare of the people.
>
> 4. The welfare of the people...[is] dependent on the closest cooperation between the district and political authorities [Afghan districts and provincial governors]...nothing must be done to weaken further that co-operation. Indeed every effort must be made to strengthen it.[37]

Shahrani, a strong proponent of what he refers to as "community based governance" in Afghanistan, echoes Bruce's conclusions. In interviews with the author, Shahrani pointed out that the local governments in towns and cities across the United States care very little about what happens in Washington, D.C., because they are more concerned with their local issues: food, housing, economic livelihood, schools, dispute settlements, and other local governance issues at the lowest level. Afghanistan's traditional local (tribal) governments address these issues adequately. When these issues are not addressed, the local leader is neither sought for help nor is he/she reelected.

Most Afghan communities at the official local government level do not receive this representation through the Durrani dominated Kabul appointed district officer (*wuluswaal*) who is not from their tribe and very often not from their ethnicity. Therefore, they rely, as they have for centuries, on their tribal community-based *shuras* or *jergahs* in which a tribal assembly of elders makes a decision by consensus to resolve the important issue at hand. This form of community-based governance leaves the Afghan government outside the decision-making loop at the local level. Frequently local communities do not witness any benefit from the national government. Instead, they see blatant corruption, cronyism and the oppressive practices of "person centered sovereignty-based" rulers at the district level and up lording over them.[38]

Shahrani suggests, moreover, that "the current vision for the Afghan government is 'madness,'" and the planned surge of 17,000 additional troops should be utilized only as a security force that promotes creating "good community-based police and the building self-defense forces within each Afghan district." He argues that Afghanistan, throughout its history has always turned to "autocratic hegemonic rulers" who appoint "key political positions" within the Afghan government such as provincial governors and district chiefs. His examination of the state of governance in Afghanistan is recorded in his *Resisting the Taliban and Talibanism in Afghanistan: Legacies of a Century of Internal Colonialism and Cold War Politics in a Buffer State*. In it, he writes "the structural dynamics of the present conflict in Afghanistan stem from "kin-based and person-centered Pashtun tribal politics" and "long simmering internal cleavages within Afghan society that pre-existed the two-decade-long war and dislocation" (referring here to the Soviet-Afghan War and the following war fought between rival anti-Soviet Mujahideen).[39]

In his "The Challenge of Post-Taliban Governance," Shahrani wrote:

> The Afghan state, like many other post-colonial states, was constituted on the basis of the old dynastic person-centered model of sovereignty in which the ruler exercised absolute power. The rulers, whether under the monarchy, Daoud's royal republic, the Khalq-Parcham Marxists, or the Mujahadeen and Taliban regimes, and currently Chairman Karzai [now President Karzai], have all attempted to rule over the country and its inhabitants as subjects rather than citizens. They have done so by relying on the use of force and maintaining/building a strong national army and police, complemented by their power to appoint and dismiss all government officials at will. Such a concentration of power has led to practices of which the outcomes have been nothing but tragic.[40]

Written in June 2003, Shahrani's warnings have been validated by the continual slippage of support for the Afghan central government in many of its districts, and the re-growth of the Taliban to fill the void of true governmental representation in rural areas. Indeed, Shahrani sees significant conflict brewing on the horizon:

> Those Afghans who hold (or aspire to) power in the capital, Kabul and wish to expand their control

over the rest of the country" ... [through Kabul appointed provincial and district officials]..."make laws, implement laws and oversee laws all under the same position which of course puts them in conflict with two other groups;" the "war-weary and impoverished masses" including "internally displaced peoples and refuges," and those "local and regional leaders and their supporters who wish to retain their autonomy from the center and ensure a significant stake in the future governance, reconstruction, and development of the country.[41]

The United States must now refocus its efforts on retaining and obtaining the disaffected Pashtun peoples (tribes) while destroying al-Qaeda and marginalizing the Taliban. This may be accomplished only through aggressive tribal engagement and partnership started at the district level and below. Improving Pashtun, as well as all other ethnic social, economic, and political structures, while slowly connecting them to the Afghan government is a long term task, but is achievable if planned appropriately and implemented in an overarching strategy; however, it must first start at the district level and below.

Reengaging the Pashtuns to Strengthen Kabul's Legitimacy

The complexities of the global and regional persistent challenges demand new and innovative ways of integrating capabilities, capacities and authorities across multiple agencies.

Civilian-led organization, even though supported by DoD, would be better able to secure local support and develop cooperative agreements.

The USG and DoD can more effectively counter violent extremist ideologies through proactive policies, greater information sharing and a broader, evolved understanding of by, with and through.

-USSOCOM Irregular Warfare Development Series Final Report, 16 February 2009[42]

The U.S. military must utilize all of the instruments of national power—diplomatic, information, military, and economic—infused with operational cultural (DIME-OC) at the strategic and operational level of war in Afghanistan and Pakistan. By doing so, the United States can destroy al-Qaeda and Taliban forces while gradually stabilizing the Pashtun tribes and people, persuading them and other Pashtuns from aligning with the enemy. This effort should resemble a weight scale approach in balancing the increased growth of stable Pashtun tribes while destroying and degrading al-Qaeda and the Taliban sanctuaries and recruitment efforts in Afghanistan and the FATA and NWFP.

Building effective local governance

Al-Qaeda and the Taliban have effectively concentrated their efforts on the disenfranchised rural Pashtun, aiming to cause a greater wedge against the first group, the ruling Durrani Pashtuns within the Afghan Karzai led government. Many Pashtuns living in the rural areas of the country are skeptical about what they hear from Kabul leadership because they have not seen tangible change in their provinces or districts in many years and have lacked qualitative representative governance at the district level and below. The Afghan government presently struggles with preventing widespread corruption throughout many of its departments and agencies. The U.S. military and the interagency programs should mentor Afghan governmental leaders, as well as provide oversight in key military and administrative billets throughout all levels of government from Kabul to the provincial and district levels, in order to attack comprehensively all the problems that the Pashtun tribes face.

The first step in ensuring that functional governance can grow and thrive is for the local tribesmen to have "buy-in," representative governance at the tribal level through their existing system. To effectively prevent the wedge from growing, the Afghan government must immediately create a constitutional mandate allowing local communities (tribes/or local village) to elect or select through their own social-political frameworks, their own leadership below the district level. Most Pashtun tribesmen, as well as

100 *Applications in Operational Culture*

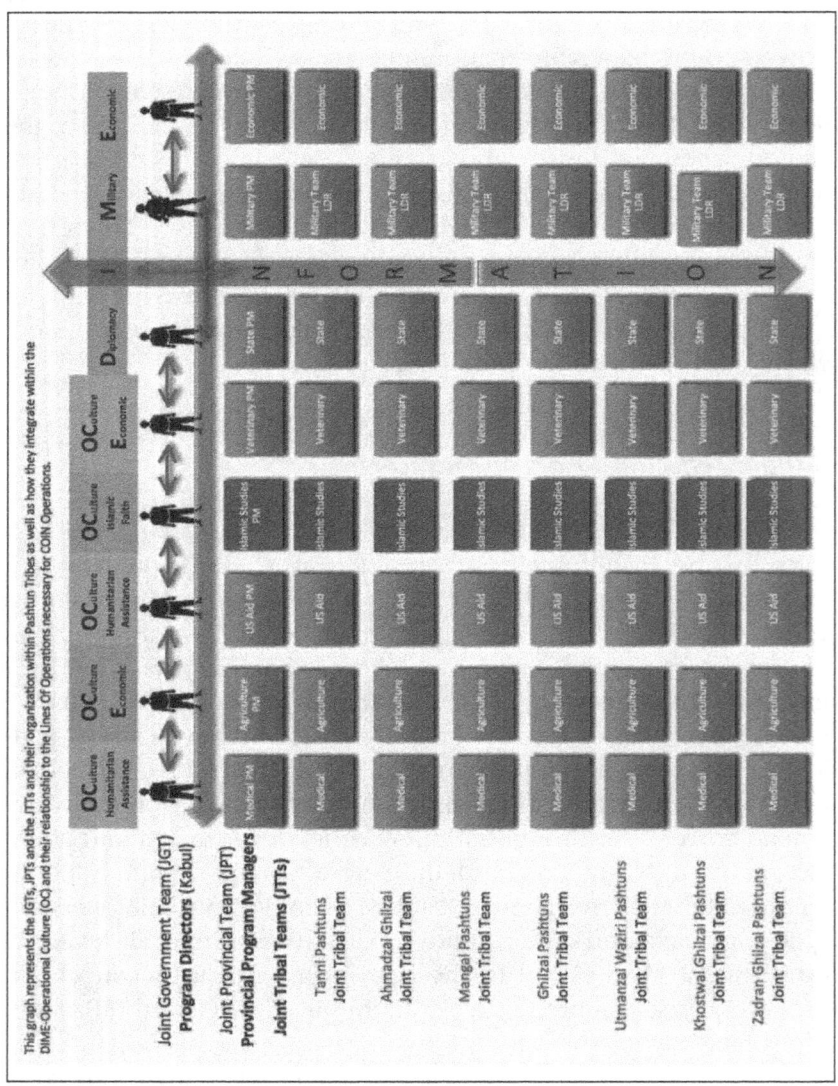

Figure 17

Example of the JGT, JPT, and JTTs produced by the author based on Khost Province, Afghanistan [note: not all tribes living in Khost are shown]. A similar structure could be implemented in Pakistan pending Pakistani approval.

other Afghan ethnicities in the rural areas of the country and the FATA and NWFP, already have a community-based government system at the tribal level that chooses a tribal leader. Many of the British military officers commonly referred to this tribal leader as "the headman."[43] It is better to work through this key individual at present in order to connect the various Pashtun rural tribes to the provincial level of government and demonstrated true buy-in and representative government. If this change is enacted now, Pashtuns at the community level would have, at a minimum, a leader of their own choosing from their tribe who is not affiliated with the Durrani-appointed district governor; therefore, building a basic level of trust and representation among the people immediately.

Ultimately, however, the only way to achieve a legitimate national government is to take the Pashtuns' tribal level government and connect it to the provincial level. In order to achieve this, the author proposes that the U.S. military and its interagency brethren form Joint Tribal Teams (JTTs).

Joint Tribal Teams

A JTT (see Figure 17 on previous page) is a joint civilian-military team, comprised of civilian and military subject matter experts, superimposed on the existing tribal or community governance framework that is being utilized in a particular district or tribe in which that JTT is assigned. The JTT mission would provide: economic stimulus; agricultural subsistence assistance and growth; veterinary support for animal husbandry production; medical aid and assistance for the entire tribe and its clans; tribal political liaison to the Afghan or Pakistani governments; and, education.

Breakdown and Relationship of the Joint Teams

At a minimum, JTTs should contain a team leader (U.S. military), medical officer (U.S. military), agricultural member (U.S. civilian agricultural specialist), U.S. Aid member, Islamic scholar (ethnic Pashtun), U.S. veterinary specialist (U.S. civilian DVM), and economic and local governance member (U.S. foreign service officer). The members of the JTT should possess the capability to assess conditions in each tribal region that they are responsible for and must have the authority to send those needs up through the appropriate Joint Provincial Teams (JPT: provincial level) and Joint Governmental Teams (JGT: Kabul-based) channels, ensuring clear lines of authority and fluid decision making processes throughout the chain-of-command.

The JTT would have within its composition an *'alim-ulma* (Islamic scholar) approved by the Afghan government who would carry with him considerable Islamic "clout" among the various tribal elders when partnering with the tribal leadership.

The Islamic scholar would work through the tribe's mullah to provide education in a culturally acceptable manner. This person would assist with the building of new mosques and schools to support Islamic education, as well as reading, writing, mathematics, social studies, and other education as necessary for the boys, and eventually girls, of the various Pashtun tribes.

The U.S. Department of State JTT member would become responsible for assessing the type of tribal governance at a particular tribe's community-based level and then link it to the provincial level through the appropriate personnel, both Afghan and U.S. This combined form of government (tribe-to-provincial government) would not look like the current system in Afghanistan (politically appointed) from Kabul to district. Instead, the arrangement and relationship would more resemble that of the Native American tribes to the U.S. government through the Bureau of Indian Affairs. Each tribal leader would be the link to the Kabul-appointed provincial governor and replace the politically appointed district chief. Implementing this change will mean that Pashtuns will have men from their ethnic, tribal background for a political voice so the tribal leader's authority will not be blocked, as it has been by Kabul political appointees and the Taliban who have done so in their push to usurp traditional Pashtun tribal governance, most commonly accomplished via the tribal *jergah*.

By having these JTTs, JPTs, and JGTs in place, the U.S. military could focus solely on providing the necessary security to each individual interagency specialist within each department, thereby allowing the interagency team member to focus on his specific mission to assist the tribe and their needs. Additionally, by implementing this system, the United States can filter through hundreds of stalled essential programs currently blocked by Afghan governmental inefficiency, cronyism, and organized crime, which exists at every level of government. The restructuring of these joint teams would break through to the heart of the problems that the tribal communities have with the Afghan national government and its local representatives. Currently, the faith in the Afghan government to provide the necessary functions of government for the Pashtuns to be secure and economically productive is missing. Over time, with constant interaction, mentoring, and oversight at the district, provincial, and governmental levels, the system of good governance can begin to take root from the bottom up eventually reaching up to the provincial level

An integral part of the overarching U.S. strategic plan focused on destroying al-Qaeda and degrading the relevance of the Taliban among the Pashtun tribesmen, must be the operational planning for the JTTs. The aim should be to grow local social, economic, and political linkages from the tribal system to the governmental system over time. In order to do this, a thorough "operational culture analysis" needs to be conducted on each specific Pashtun tribe to examine the needs that al-Qaeda and the Taliban cannot meet. The U.S. military, in partnership with the JTTs, JPTs, and JGTs will require significant assistance from civilian specialists and government analysts who specialize in the specific tribal region to include tribal decent, history of conflicts with other tribes, and all affiliations both politically and economically, down to the tribe and clan level of detail.

Operational planning should avoid a purely military solution during initial phases of the new strategy. The use of military force should be relegated to designated phases or during critical phases of the overarching strategy, and in most cases, should only be focused on securing districts to enable the local tribes to protect themselves. The U.S. military presence should be a means to build local community security and should not be utilized as a means to arm militias to fight al-Qaeda on behalf of the U.S. and Afghan governments. This point should be clearly distinguished. Any U.S. military force that is used should be tailored for that specific situation or mission and complement the overall regional and tribal objectives concurrent with the JTTs', JPTs', and JGTs' objectives and the operational commander's intent. Irrespective of level (tribal, provincial, government), the joint teams should consider the following six points in their operational planning:

> 1. Humanitarian relief operations similar to those employed at the Marine Expeditionary Unit (MEU) level, on-call, for any area within the region that

the JTTs are operating throughout all key phases. Humanitarian aid should contain the capacity to be funneled through local tribal chiefs and district level administrators when being distributed within that particular tribe's region.

2. Respect for Pashtun tribal martial strength, as well as honoring the insight and wisdom of the tribal leaders in how to use it wisely. Pashtuns decide to commit or not commit their tribes to conflicts based on discussion in lengthy *jergahs* that examine what is to be gained in fighting for a particular cause or leader. The JTT leaders must amplify the need to work together instead of fighting one another. The tribe's military value should be acknowledged openly and their fighting ability should be reiterated and respected throughout all dealings with them. Fighting and warfare is a vital part of Pashtun tribal culture; approaching them in any other manner would not be normal.

3. Mitigate damage to tribes that live among al-Qaeda and Taliban forces currently. JTTs should utilize considerable leveraging and negotiations between friendly tribes working with the JTTs and those supporting al-Qaeda and the Taliban, with the goal of splitting tribes away from al-Qaeda and the Taliban rather than fighting them along with al-Qaeda and the Taliban.

4. All operational phases should be event-driven and designed with the flexibility to lengthen or shorten phases within specific tribal areas when required. Some tribes will require more time than others due to al-Qaeda and Taliban influence, numbers in population (sub-tribes and clans in remote areas), difficulty in reaching them (weather/terrain), hasty *jergahs* to mediate deals, or ongoing blood feuds between tribes that are not necessarily related to providing support to al-Qaeda or Taliban. Every effort must be made to contact all reachable tribes.

5. A thorough cultural ethnographic study and intelligence report of all the tribal areas in the FATA and NWFP. The JTTs should determine and prioritize the focus of effort. They should work with the most enemy influenced tribe first; or last, or a combination of the two—listing the tribal regions as green, yellow, and red tribal zones.

- Green zones should be those areas where there is no perceived al-Qaeda and Taliban threat or threat of partnering between the Pashtuns and al-Qaeda and Taliban.
- Yellow zones should be identified as those areas where contact was made and the tribal *jergah* decided that the tribe would not work with the JTT, but would stay passive on the sidelines.
- Red zones should be those areas to which the JTTs could not gain access due to al-Qaeda and Taliban dominance and/or tribal *jergah* hostility. Note, in some cases, tribes will restrain from contact to protect their villages from retribution or to protect other tribal interests such as an ongoing war with a bordering tribe, drug trafficking, or protection of other unknown interests.

6. All joint team staff should contain native Pashtuns from the various tribes who can work with the JTTs to find inroads to key Pashtun tribal leaders and tribesmen to request and schedule tribal *jergahs* under the flag of Pashtunwali, to provide the safe access for JTT-tribal dialog.

Pashtuns and Islam

Considering the Pathan's [Pashtun] eminently material form of faith, his confidence in saints and shrines, prophets and priests, [Sufi Islam] prayers and pilgrimages, it is not a little curious that religious reformers directly opposed to all this should from time to time have obtained such influence on the border [similar to al-Qaeda-Taliban today].

-Edward E. Oliver, *Across The Border* (1890)[44]

Today, the U.S. military struggles with the same realization as that of the British: The Pashtuns hold their tribe as high, and in some cases, higher than the level of Islamic cohesiveness among themselves. Some U.S. forces in Afghanistan today have a misconception that the Pashtuns are united solely by the Islamic precepts derived from the *Ummah* (Islamic community) or their specific Sufi, Shia or Wahabi-ist denomination of Islam. This is not true. What is evident is that the Taliban today are providing mullahs of their own choosing and training them to preach to the Pashtun tribesmen and employing them as key information and propaganda nodes throughout the tribal belt. Meanwhile, the Afghan government's voice (from Kabul) and their version of Islam is being drowned out by Wahabi extremist ideology. Pashtuns are becoming increasingly swayed by Taliban-al-Qaeda information operations from the *minbar* (pulpit) and moving farther away from their reliance on older tribal governing bodies (*jergahs*) that once held a higher level of authority in the decision making process for their entire social structure (clan and tribe). In *Frontier of Faith*, a nuanced and comprehensive study of Islam and its role in the Pashtun tribal order, Sana Haroon notes:

> These understandings of history and society, reliant on oral and transcribed genealogies which described homogeneous clans and tribes of communities descended from a common mythical forefather Qais Abdur Rashid, cannot accommodate the membership of mullahs in Pakhtun society as their participation was functional and not genealogical. By such reasoning, reinforced by the fact that organization of the *sisila* and the *shajarah* [Sufi Saints] were rooted in separate myths of lineage, systems of representation and sources of patronage, mullahs have been understood to have been mere 'clients' of the tribal system and incidental to its functioning. Yet in the space of the

non-administered Tribal Areas [FATA-NWFP], religious practice, deeply influenced by Naqshbandiyya-Mujaddidiyya [the "silent Sufi" who practiced silent meditation] revivalism and the village and community-based activities of the mullahs, gives little evidence of real distinctions between tribal social organisation and motivation, and the activities of the mullahs. Mullahs participated in village-based community living: trading, interacting and intermarrying within the clan unit. In almost all cases mullahs were ethnically Pakhtun, and in many cases were originally from the clan that they served.[45]

Haroon's analysis suggests that a specific Pashtun tribal mullah who adheres to the Taliban-al-Qaeda-supported Wahabi Islam takes on a central role in the tribe or clan as well as in other important social and political functions. Not only do they provide religious instruction, but they also perform marriages, other tribal ceremonies, and offer prayers before and after tribal *shuras*. One can see how complex the situation gets, and to what degree operational culture planning must occur, before trying to "drown-out" the radical doctrine from each Pashtun tribe's Mullah, which he uses to recruit young Pashtun men who fight against American, Afghan, and Coalition forces in these regions. This issue becomes even more complicated when some elders within the tribe might be at odds with the mullah's teachings (non-Sufi), since the elders still need the Mullah to perform vital social functions (i.e., marriage ceremonies for their sons and daughters), which are tied to complex economic and political relationships within the clan and tribe. The social pressures on these tribal elders is immense and many would not want to be seen as falling out of favor with the one spiritual leader (holy man) within the tribal community.

In 1890, Briton Edward Emerson Oliver described the Wahabi influence in the tribal region. He wrote:

> There is certainly less difference between the ritualistic Pathan [Pashtun] and the puritanical Wahabi, the latter very often a despised Hindustani to boot.

> Yet in spite of the long record of mischief, and the troubles which this mischief has led to, colonies of these and other religious adventures of all sorts have been sheltered, and, so long as they refrain from interfering with Pathan custom, have been protected, and even cherished.[46]

Oliver's perplexity as to why the predominately-Sufi Pashtuns would house such a different sect of Muslims has resonance today for U.S. forces who are uncertain of the degree of loyalty within the present Pashtun tribe-Taliban-al-Qaeda alliance.

Elsewhere, in his book, Oliver describes a situation that is not uncommon for those living amongst the Pashtuns, the story of a "formidable apostle of insurrection," Sayud Ahmad of Bareilly. Sayud Ahmad was a "student of Arabic under the learned doctors of Delhi," and was adept at fostering dissension among the various tribes against the infidels and Sikhs in the Punjab (similar to Bin-Laden today). Ultimately, Sayud was expelled after his "ill advised effort to reform the Pathan [Pashtun] marriage customs, which was really an attempt to provide wives for his own hindustanis." The Pashtuns on their own, "simultaneously massacred his [Sayud's] agents, and in one hour, the hour of evening prayer–they were murdered by the tribesmen [Pashtun] almost to a man."[47] This story is not unsimilar to al-Qaeda-in-Iraq's (AQI) efforts to marry the daughters of Sunni tribesmen in Fallujah, 2006—although in this case there was no massacre.[48]

The Afghan government, with the assistance of U.S. security forces and information operations, must begin to plan and build re-formed Afghan and Pakistan government-approved Islamic educational systems managed by moderate mullahs with new curricula. This new education program should be implemented within each remote tribe and clan below the district level. These efforts should be connected to the government through an 'Alim-Ulama from the highest level of Islamic schooling down to the tribal level. The 'Alim-Ulama should facilitate Islamic education (Quran recitation and study) as well as other key components of secular education: reading, language, writing, mathematics, science, and social studies throughout the border regions of eastern Afghanistan as well as within the FATA and NWFP.

Islamic education and training for mullahs and Islamic students (*Malawais*) was an extremely important part of Pashtun tribal culture during the British experience, as it still is today. Many tribes within the FATA and NWFP desire religious education for their young men. It is the only formal education most will receive in their lifetime. *Madrassas* fulfill the extensive demands for religious training of young Pashtun men and many other social functions as mentioned. The U.S. and Afghan governments cannot diminish the strength of the radical Wahabi-sponsored *Madrassas* unless it has a plan to replace them with a better alternative for the people. Replacing Wahabist *Madrassas* is a long-term effort that must be undertaken in order to stabilize the region and diminish al-Qaeda's strength and credibility.

It is important to note that the mosques within each Pashtun village are social loci. Pashtuns that witness beautiful mosques that are built through government funding and support local social [Islamic spiritual] and economic growth will no longer accept the al-Qaeda and Taliban mantra that the U.S. and the apostate rulers (Afghanistan and Pakistan) are in the region to destroy Islam. Additionally, if local tribesmen build the mosques for their own village with government funding, it stimulates their local economy, builds ties with the Afghan and Pakistani governments on religious commonality lines, shows true relationship building, and provides other avenues for social participation and connection among the tribe. The '*Alim-Ulma* JTT member will also be able to introduce the other JTT members at future *jergahs* to begin the initial discussions on how to select and prioritize agricultural, health services, economic, and educational projects. Finally, if al-Qaeda or the Taliban destroy the mosques that are being built or threaten those who participate in constructing or worshiping in the mosques, al-Qaeda or the Taliban could lose credibility among the Pashtun tribesmen. In fact as the fate of Sayud Ahmad of Bareilly suggests, it would be dangerous to be viewed in a negative manner by the Pashtun people within a particular village or region.

Partnering With Pakistan

The mountainous borderlands where Afghanistan meets Pakistan have been described as a Grand Central Station for Islamic terror-

Applications in Operational Culture

ists, a place where militants come and go and the Taliban trains its fighters.
-"Pashtunistan Holds Key to Obama Mission," *The Guardian*, February 2009[49]

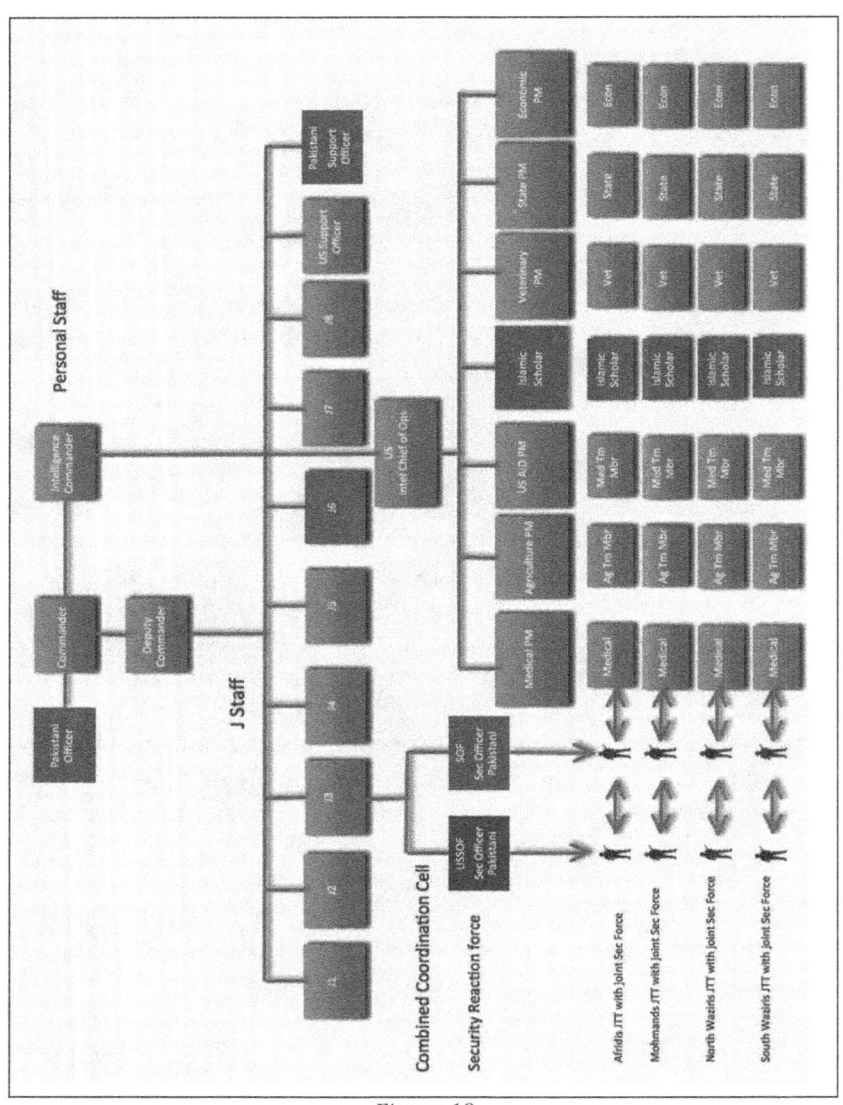

Figure 18

Example of a Joint U.S.-Pakistani Command Structure to support Pakistani JTTs

There cannot be a stable Afghanistan with an enemy sanctuary in the FATA and NWFP. The FATA and NWFP are very different on many levels from Afghanistan. The Pashtuns residing in these regions such as the Karlanri are made up of diverse sub-tribes: Afridis, Daurs, Jadrans, Ketrans, Mahsuds, Mohmands, Waziris (North and South), and many other sub-tribes and clans[50] that have different social, economic, and political structures than those Pashtuns living in lower regions of Afghanistan (Durrani). To build effective relationships with them, the United States must get out of the cities and off the roads and enter into the foothills and mountains of the Pamir Range, Samana Range, Safed Koh Range, Waziristan Hills, and south into the Sulaiman Range and meet them.

The U.S. military is not capable of conducting prolonged combat operations in these regions simply due to the sheer number of Pashtun tribesmen that inhabit the region, the difficulty of providing logistical support, and the Pakistani government's reluctance to allow a large U.S. footprint within its country. As Pashtuns living in these regions do not recognize the governments of Pakistan and Afghanistan, the issue of meeting Pashtun tribesman and persuading them to cease support for the Taliban and al-Qaeda, although problematic, must be considered. Therefore, the U.S., through the highest diplomatic channels possible, must request limited access for the JTTs and a small joint U.S.-Pakistani security force partnership to enable the JTTs to have Pakistani-supervised access to this region. The purpose of the joint U.S. security forces would be to provide security for the JTTs and allow them to gain access to the Pashtun tribes, under the flag of *Pashtunwali*, residing in this area. If allowed, all aforementioned recommendations of JTT operations and missions would be applicable, save some particular aspects related to Pakistani military consent and security measures.

If the JTTs are able to operate in the FATA and NWFP, enemy attacks on the western side (Afghanistan) of the Durand Line would drop significantly due to the threat that they would pose to al-Qaeda and the Taliban, in their current safe-haven, who need to prevent tribal alliances from forming against them. By eliminating Pashtun support from al-Qaeda and the Taliban, the United States can deny these organizations sanctuary. In order to eliminate Pashtun support, an "awakening" must occur among Pashtun tribes. This awakening must be different from the one that occurred among the Sunnis in Iraq, due to the different cultural, religious, and territorial issues related to the tribes living in the FATA and NWFP. The U.S. military must be careful in making too many assumptions about Pashtun tribes based upon the Iraqi model.

The United States must begin an earnest partnering effort with Pakistan that is built upon a shared relationship at all levels that respects and accepts the concerns of both nations. Therefore, operational planning must include Pakistani government "buy-in" and must be demonstrated in a joint U.S.-Pakistani command structure that includes the selection of JTTs with Pakistani participants, especially within the *Alim-Ulama* positions and any issues concerning Pakistani government liaison within the JTTs. This structure should physically show Pakistani participation by opening up key military and civilian staff billets to the Pakistanis and placing them within the joint command. The joint Pakistan-U.S. command should focus on planning that destroys al-Qaeda, marginalizes the Taliban, and dissuades the Pashtun tribes from embracing the ideologies of both of these entities, while at the same time improving the social and economic conditions of the various Pashtun tribes living in the FATA and NWFP. The joint command should also plan ways and means to stimulate and support the region's tribal economy within the FATA and NWFP through joint U.S.-Pakistani monetary efforts. The primary responsibility of Pakistanis on the JTTs will be to provide Pakistani-to-Pashtun tribal governmental liaison personnel to the teams, as well as to identify those areas that will facilitate the gathering of tribal elders at a *jergah*.

The Pakistani military must also agree up front with the size and scope of the U.S. advisor mission and must clearly understand and accept their partnering role with Pakistan forces. The Pakistan Inter-Services Intelligence (ISI) must be planned into every phase of the joint U.S.-Pakistani operation. ISI members have an historical affiliation (working and familial) with al-Qaeda and the Taliban on many levels. Complete understanding and agreements must be clearly defined and strict security measures placed to restrict critical information that might compromise both U.S. and Pakistani efforts in combating al-Qaeda. Operational planning between Pakistan and U.S. forces should focus on strategies and courses of action that will work within the known geopolitical complexities and friction-points tied to the FATA and NWFP, and must be able to quickly negate their destabilizing effects on the sovereignty and political structure of Pakistan, while keeping a strong Pakistani border intact. Operational planning should forbid the involvement of Afghanistan on any matter across the eastern side of the border within Pakistan and Afghan forces should focus solely on the western side in JTT with joint U.S.-Afghan operations against al-Qaeda and Taliban forces operating there.

The overall operational commander of the Joint Pakistani and U.S. command should have command authority of both western and eastern sides of the border (Durand Line) and freedom to move his forces and assets when and where needed after careful consultation with his Pakistan counterpart. Operational planning should employ a U.S. Intelligence Commander who is solely responsible for battlespace on both sides of the Durand Line. This U.S. Intelligence Commander must be closely linked to the Operational Commander and must be given freedom of movement to place key facilitating personnel in the right place in order to assist joint U.S. Pakistani forces. The U.S. contingent should be able to train a select group of Pakistani military to be used as a joint security team, which will be joined with U.S. advisory forces to provide security for the JTTs when they operate within the FATA and NWFP. These JTT security teams should train together during the preparation phase of all operations for each region into which the JTTs would deploy.

Conclusion

The U.S. strategy in Afghanistan is at a crossroads. U.S. strategy can either continue with the current policy of employing U.S. forces to support a government that is illegitimate in the eyes of many Pashtun tribes, or the United States can focus on winning over disaffected tribes by using an intense tribal engagement: a program consisting of economic, social, and political investments from the bottom up, while slowly building the Afghan government from the top down. By utilizing this integrated approach, Afghanistan will gradually attain a form of representative-governance with buy-in and legitimacy from all of its people, not just Pashtuns.

The process to achieve this integrated approach may require several years, if not decades. If U.S. strategy continues to focus on building an Afghan central government that is dominated by Durrani Pashtuns from the top down, while excluding tribal relationship building and partnering with the Ghilzai, Karlanri, and other rural Pashtuns and ethnicities from the bottom up, the legitimacy of the new Afghan government will never be attained. The Taliban leaders will continue to exploit their geographical and social advantages, while actively recruiting future generations of Pashtun tribesmen throughout the eastern and southern provinces of Afghanistan, the FATA, and the NWFP.

By focusing on warfighting concepts, the historical record, operational art, operational culture, and a new joint interagency approach, the author proposes a bold unconventional path to achieve success against an enemy that has fought the U.S. irregularly for eight years. Through careful research and study, interviews with Pashtun tribal leaders, and experienced gained serving along side Pashtuns as a military advisor from 2003 to 2005, the author employed, with significant success, some of the operational culture methods described in this article. Continued at the strategic and operational level and in a much broader scale, these methods and approaches, which are built on relationships and trust, will eventually allow the U.S. to undermine al-Qaeda and diminish the relevance of the Taliban among the Pashtun, paving the way for a safer, more representatively governed Afghanistan.

Notes

[1] Megan K. Stack, "The Other Afghan War," *Los Angeles Times*, 23 November 2008.

[2] Thomas H. Johnson and M. Chris Mason, "All Counterinsurgency Is Local," *Atlantic Monthly*, October 2008.

[3] Anna Mulrine, "U.S. General: Many in Afghanistan 'Don't Feel Secure,'" *U.S. News & World Report*, 19 January 2009.

[4] U.S. Department of Defense, *Joint Publication 5-0: Joint Operation Planning*, 26 December 2006, v-8.

[5] LtCol Antonio J. Echevarria II, *Clausewitz's Center Of Gravity: Changing Our Warfighting Doctrine-Again!* (Carlisle, PA: Strategic Studies Institute, U.S. Army War College, 2002), 16.

[6] Ibid., 13.

[7] Russian General Staff, *The Soviet-Afghan War, How a Superpower Fought and Lost,* trans. and ed. Lester W. Grau and Michael A. Gress (Lawrence: University Press of Kansas, 2002), 25.

[8] Ibid., 26.

[9] HQ Department of the Army, Washington, DC; HQ Marine Corps Combat Development Command, Department of the Navy; HQ United States Marine Corps,

Washington, DC, *FM 3-24 (MCWP 3-33.5) Counterinsurgency* (Washington, DC: U.S. Department of Defense, 2006), 1-21.

[10] Nathanial C. Fick and Vikram J. Singh, "Winning the Battle, Losing the Faith," *International Herald Tribune*, 6 October 2008.

[11] M. Nazif Shahrani, "The Challenge of Post-Taliban Governance," *ISIM Newsletter*, 12 June 2003.

[12] Fick and Singh, "Winning the Battle."

[13] Thomas J. Barfield, "Weapons of the Not So Weak in Afghanistan: Pashtun Agrarian Structure and Tribal Organization for Times of War and Peace," *Hinterlands, Frontiers, Cities and States: Transactions and Identities* (Agrarian Studies Colloquium Series), 23 February 2007, 1.

[14] Lawrence James, Raj: *The Making and Unmaking of British India* (New York: St Martin's Griffin, 1997), 394.

[15] Olaf Caroe, *The Pathans* (London, UK: Oxford University Press, 1958), 11.

[16] Bernt Glatzer, "The Pashtun Tribal Systems," in Deepak K. Behera and Georg Pfeffer, eds., *Concept of Tribal Society* (New Delhi, India: Concept Publishers, 2002), 5.

[17] Thomas H. Johnson and M. Chris Mason, "No Sign until the Burst of Fire: Understanding the Pakistan-Afghanistan Frontier," *International Security* 32 (Spring 2008): 51.

[18] Glatzer, "Pashtun Tribal Systems," 7.

[19] Johnson and Mason, "No Sign until the Burst of Fire," 51.

[20] Shahrani, "Challenge of Post-Taliban Governance."

[21] Frank A. Clements, *Conflict in Afghanistan: A Historical Encyclopedia* (Washington, DC: ABC-CLIO, 2003), 81-82.

[22] Barfield, "Weapons of the Not so Weak," 11.

[23] Ibid.

[24] Ibid.

[25] Johnson and Mason, "No Sign until the Burst of Fire," 59.

[26] Carter Malkasian and Jerry Meyerle, *A Brief History of the War in Southern Afghanistan* (Quantico, VA: U.S Marine Corps Intelligence Activity, 2008), 10.

[27] Ibid.

[28] Anthony Loyd, "British Frontier Corps Veteran Recalls Fighting Pashtun Tribesmen," *London Times*, 2 January 2009.

[29] Mountstuart Elphinstone, *An Account of Caubul, and its Dependencies, in Persia, Tartary, and India; comprising a View of the Afghan Nation, and a History of the Dooraunee Monarchy* (London: Richard Bentley, 1839), 210.

[30] Ibid.

[31] Ibid.

[32] James Atkinson, *The Expedition Into Afghanistan: Notes and Sketches Descrip-*

tions of the Country, Contained in a Personal Narrative during the Campaign of 1839 & 1840, up to the Surrender of Dost Mahomed Khan (London: William H. Allen, 1842), 4.

[33] BGen Sir Percy Sykes, *A History of Afghanistan* (London: MacMillan, 1940).

[34] "Cotton, Sir Sydney John (1792-1874)," in *Australian Dictionary of Biography*, online edition (http://adbonline.anu.edu.au/biogs/A010240b.htm).

[35] LtGen Sir Sydney Cotton, *Nine Years on the North-West Frontier of India, from 1854 to 1863* (London: Richard Bentley, 1868), 13.

[36] LtCol C. E. Bruce, *Waziristan 1936-1937: The Problems of the North-West Frontiers of India and Their Solutions* (London: Aldershot Gale Polden, 1938), 73.

[37] Ibid.

[38] M. Nazif Shahrani, Indiana University, discussion with author, February 2009.

[39] M. Nazif Shahrani, *Resisting the Taliban and Talibanism in Afghanistan: Legacies of a Century of Internal Colonialism and Cold War Politics in a Buffer State* (Bloomington: Indiana University Press, 2008).

[40] Shahrani, "Challenge of Post-Taliban Governance," 22.

[41] Ibid.

[42] "USSOCOM Irregular Warfare Development Series Final Report," 16 February 2009.

[43] Bruce, *Waziristan 1936-1937*, 73.

[44] Edward E. Oliver, *Across The Border* (London: Chapman and Hall, 1890), 286.

[45] Sana Haroon, *Frontier of Faith: Islam in the Indo-Afghan Borderland* (New York: Columbia University Press, 2007), 66.

[46] Oliver, *Across The Border*, 286.

[47] Ibid.

[48] Alissa J. Rubin, "Sunni Sheik Who Backed U.S. in Iraq Is Killed," *New York Times*, 14 September 2007.

[49] Jason Burke, Yama Omid, Paul Harris, Saeed Shah, and Gethin Chamberlain, "'Pashtunistan' Holds Key to Obama Mission," *The Guardian*, 15 February 2009.

[50] Peter R. Lavoy, *Pakistan's Strategic Culture* (Fort Belvoir, VA: Defense Threat Reduction Agency, Advanced Systems and Concepts Office, 2006), 9.

Chapter 5

The Application of Cultural Military Education for 2025

Major Robert T. Castro, USMC

Major Robert T. Castro is a Marine supply officer with more than 22 years of service. During his career, he has been assigned to 3d Battalion, 1st Marines; on a western Pacific deployment with the 15th Marine Expeditionary Unit; with 1st Marine Corps Recruiting District as a contracting officer; and II Marine Expeditionary Force Headquarters Group, deployed twice with the unit to Operation Iraqi Freedom. During his second deployment with the MEF, he held the operational billet of Mayor of Fallujah. As the mayor, he had oversight of three forward operating bases and was responsible for reconstruction assistance to build and repair city infrastructure. He also acted as a participant in the city council meetings.

Castro's follow-on assignment sent him to the Institute for Defense Analyses, Joint Advanced Warfighting Division, where he was the team lead for two projects, one a study on Iraqi tribal environment and structure for Multi National Forces-Iraq, and the other a project on disarmament, demobilization, and reintegration. Both projects required extensive field interviews and cultural studies within Iraq.

Castro graduated with marketing degrees from Concord University in 1988 and earned a bachelor of science degree from the George Washington University in 1992. He is a recent graduate of the Marine Corps School of Advanced Warfighting, attaining a masters in operational studies.

And when people are entering upon a war they do things the wrong way around. Action comes first, and it is only when they have already suffered that they begin to think.

-Thucydides, Book 1, Par 78

If you know the enemy and know yourself, you need not fear the result of a hundred battles.

-Sun Tzu, *Art of War*

In 1918, the British Army fought in Iraq against the local tribes to control the cities of Fallujah and Baghdad after the Ottoman Empire had fallen. The British experienced large troop losses and an inability to win over the populace; one British officer bitterly described Iraq as a place where "the people are animals and savages in an uncouth culture." If one forwards to the Baghdad of 2007, one finds the U.S. military experiencing the same problems the British had. This is due to the U.S. military's inability to understand who it is fighting, who it is operating amongst, and in what cultural environment it is acting. During the British, French, and American colonial period, the social scientific field of anthropology became an essential tool to understand indigenous cultures. One can distill many enduring cultural lessons from the British colonial and the American counterinsurgency experience. These enduring lessons point to a central idea: cultural expertise and thorough knowledge of the populace are essential to success in irregular warfare (IW) conflicts. As IW conflicts will continue to offer missions for the military, it is essential that the Marine Corps and the Department of Defense (DoD) integrate cultural anthropology into military planning and training today for effective warriors in the future

Cultural Anthropology and Irregular Warfare Operations

The object of DoD's Quadrennial Roles and Missions Report is to project the future planning of DoD, with its core function as defining the required missions. The January 2009 report lists irregular warfare as a core competency and mission area for the military to focus on for the future. IW is defined by the DoD Joint Operating Concept as:

> A violent struggle among state and non-state actors for legitimacy and influence over the relevant populations. IW favors indirect and asymmetric approaches, though it may employ the full range of

military and other capabilities, in order to erode an adversary's power, influence, and will.[1]

The range of operations and activities that count as a part of IW include insurgency, counterinsurgency, terrorism, counterterrorism, stabilization, security, transition and reconstruction operations (SSTRO), and psychological and information operations.

Irregular warfare operations differ from conventional combat operations in a number of fundamental ways. One of the most significant differences—one that perhaps causes the greatest difficulty for U.S. military planners—concerns the battle terrain. Unlike conventional combat operations with their emphasis on destruction of enemy military capabilities and control of physical terrain, the primary factor for success of IW conflicts depends on the support—or at least the unobtrusive acquiescence—of the civilian population. Human terrain is more important than physical terrain in IW operations.

IW operations apply to a wide range of conflicts, but the most common form of IW operation that current U.S armed forces confront is counterinsurgency. Despite the surge of American interest in counterinsurgency theory and practice in the wake of Operation Iraqi Freedom, making its popularity appear recent, counterinsurgency has always been a frequent topic of military study. In the latter half of the 20th century, Cold War-era communist and anti-colonial insurgencies spurred counterinsurgency study to much greater detail and frequency. Almost without exception, those who have planned, lead, fought, or researched insurgencies and counterinsurgencies affirm the importance of winning the civilian population in this form of irregular warfare. A few examples below help to elucidate this essential point.

One of the most frequently cited military theorists of the 20th century is Mao Zedong, who led the Chinese Community Party to victory through an insurgency strategy. In his now-famous and often-quoted analogy, the insurgent lives among the people just as a fish swims in the water.[2] The people are the human terrain that provides sustenance and freedom of action for the insurgent. Many counterinsurgency theorists have used Mao's characterization of the nature of insurgency warfare to proscribe counterinsurgency techniques for draining the water, poisoning the water, or simply changing the characteristics of the water to make it inhospitable for fish.

The French military officer David Galula observed British counterinsurgency efforts in Malaya during the 1940s through the 1960s. In his highly regarded book, *Counterinsurgency Warfare: Theory and Practice*,[3] he noted that the insurgent has inherent weaknesses and disadvantages that would make it foolish to attack his enemy in a conventional manner. The insurgent's goal was not to destroy his enemy's forces or conquer territory, but rather to force his opponent to fight at a time and place of the insurgent's choosing:

> The population represents this new ground. If the insurgent manages to dissociate the population from the counterinsurgent, to control it physically, to get its active support, he will win the war because, in the final analysis, the exercise of political power depends on the tacit or explicit agreement of the population or, at worst, on its submissiveness.[4]

Where Mao emphasizes the importance of popular support for the insurgent, Galula does the same for the counterinsurgent. Galula argues that while it may be relatively simple to mass military and police forces and clean an area of insurgent forces and support cells, keeping that area clean after the counterinsurgent forces have moved on is the essential and more difficult challenge. Counterinsurgents usually cannot afford to maintain a heavy presence in all the areas, thus explaining the importance of winning the populace to isolate the insurgents among them. He writes, "It is impossible to prevent the return of the guerilla units and the rebuilding of the political cells unless the population cooperates. The population, therefore, becomes the objective for the counterinsurgent as it was for his enemy."[5]

Robert Thompson, a British diplomat with administrative experience in the Malaysian counterinsurgency campaign from 1950-1957 and advisory experience in Vietnam in the early 1960s, echoes this counterinsurgency tenet in his manual, *Defeating Communist Insurgency: The Lessons of Malaya and Vietnam*. Although the needs of the warring parties—and therefore, the exact importance of the population's role—vary from one conflict to another, the basic premise remains the same: The support of the people whether tacit or explicit, is a necessary element of success. Hence, he posits, "An insurgent movement is a war for the people."[6]

Julian Paget, a British officer with experience in the United Kingdom's extensive counterinsurgency operations in the 1950s and 1960s, also describes the seminal importance of possessing the population's support in IW operations. He writes, "The support of the local population is an important factor for both sides in any counterinsurgency campaign." Moreover, he argues that the long-term objective of securing the loyalty of the people is "the ultimate aim of both sides, and is a predominant factor in all their thinking."[7] Paget, however, also introduces another important element not often addressed in other counterinsurgency writings. Counterinsurgents must "understand their [insurgents'] minds, their mentality, and their motives."[8] This additional consideration—that the insurgent enemy has his own motivations and perceptions that must be addressed and or countered—only further adds to the importance and relevance of studying culture for IW conflicts.

Therefore, if the population is the most important element in counterinsurgency, then a complete understanding of that element is essential. Meanwhile, understanding the insurgents' motivations and goals falls in line with the traditional military dictum of "know thy enemy." These two concepts are not new in the conduct of military operations, and thus it is not surprising that they occupy prominent positions within our military's latest doctrine on counterinsurgency. The U.S. Army and Marine Corps recently published counterinsurgency manual, FM 3-24, which frequently emphasizes the importance of understanding the population:

> Understanding the attitudes and perceptions of the society's groups is very important to understanding the threat. It is important to know how the population perceives the insurgents, the host nation, and U.S. forces. In addition, HN and insurgent perceptions of one another and of U.S. forces are also very important.[9]

Learning about the ideology and motivations of the people allows the U.S. military to better identify and exploit divisions between themselves and the insurgents. This fundamental truth has been recognized in the 20th century works previously mentioned;

however, it was not discussed in depth until the publication of FM 3-24, which is arguably the most complete counterinsurgency publication ever developed by the U.S. Army and Marine Corps.

Know Thy Enemy (And Thy Friends)

The failure to understand foreign cultures has often been a contributing factor in American diplomatic and military failures over the past 150 years. Lack of cultural knowledge within the national security establishment, limited guidance for nation-building and educational programs, and poor socio-cultural awareness development and planning are the root factors for security challenges of today and tomorrow. While the U.S. military has some organizations that address cultural issues at both the operational and tactical levels—Psychological Operations (PSYOPS), Civil Affairs, and Special Forces address the cultural issues in Foreign Internal Defense (FID) missions.

Since anthropology focuses on the study of societies and cultures, several anthropological techniques exist to help military planners gain understanding of the populace. One of anthropology's central tenets is the importance of utilizing cultural relativism when viewing other societies: understanding society from within its own frameworks rather than from without. Anthropology also emphasizes the importance of using historical research, fieldwork, and ethnographic surveys to understand the subject population. Since anthropology uses culturally sensitive methods, unlike other social sciences or military intelligence, it is uniquely suited as a means to obtain knowledge, insight, and understanding of the human terrain within an area of operations. Therefore, anthropological approaches must be incorporated into the planning and conduct of low-intensity military operations.

Although anthropology provides valuable models for understanding other cultures, only one field of the discipline is directly relevant for military planning. Anthropology is divided into four distinct fields and subfields: physical anthropology, archaeology, sociolinguistics, and cultural anthropology. Cultural anthropology, also known as socio-cultural anthropology, is the holistic study of humanity through long-term, investigative, intensive field studies to better understand the culture and social organization of a particular group of people.[10] Cultural anthropology examines many shared areas of interest for military operations: language, economic and political organizations, kinship, gender relations, religion, mythology, and symbolism. These areas of inquiry provide models for understanding a society's social networks and behavior, kinship patterns, law, and overall outlook on life. These efforts fit within anthropology's central objective of understanding the, "system of inherited conceptions expressed in symbolic forms by means of which people communicate, perpetuate and develop their knowledge about and attitudes toward life" (essentially a society's cultural framework).[11] Culture is crucial to the analysis of human behavior, for:

Without the direction of cultural norms, the system of meaningful symbols, human behavior would be absolutely ungoverned. It would turn into chaos and pandemonium of meaningless actions and emotional outbursts. Culture, in its cumulative totality of patterns of this type, is not a mere adornment of humanity's existence, but an essential condition, and as such the basis of its uniqueness.[12]

Anthropologists achieve such detailed analysis through ethnography, which requires developing rapport, selecting informants, transcribing texts, taking genealogies, mapping fields, and keeping diaries to further understand the population. Ethnography is an unobtrusive and responsive method to understand culture—a sharp contrast to the DoD and military approach, which uses checklists and assumptions without investing the same level of detailed data collection.

In the October 2004 issue of *Proceedings*, Major General Robert Scales Jr., USA (Ret), noted that during the *cultural* phase of the Iraq war, "intimate knowledge of the enemy's motivation, intent, will, tactical method, and cultural environment has proved to be far more important for success than the deployment of smart bombs, unmanned aircraft, and expansive bandwidth."[13] He further noted that success "rests with the ability of leaders to think and adapt faster than the enemy and of soldiers to thrive in an environment of uncertainly, ambiguity, and unfamiliar cultural circumstances."[14] The leadership must get inside the enemy's decision loop, and soldiers must be trained in more than tactics, techniques, and procedures. Successful cultural training should assist top-level decision makers to achieve superior decision making at the strategic level, and help soldiers on the ground develop more effective means of winning the trust and confidence of the local populace.

When placed in the context of military application, cultural awareness not only allows military personnel to operate effectively because their, "ability to recognize and understand the effects of culture on people's values and behaviors" is heightened,[15] but also because it serves as a force multiplier. Cultural awareness is not only important for use against an insurgent enemy and the indigenous population. With U.S. armed forces operating more regularly with coalition and allied forces, having a basic understanding of the cultural differences and perspectives with our partners is crucial to mission success. For instance, when a coalition partner offers a confident "no problem" in an operational planning session, he may very well be answering the wrong question. As Congressman Ike Skelton notes in his essay, "You're Not From Around Here, Are You?": "We must first begin by understanding the attitudes of host nation and coalition partners. It is only then that the military can start understanding other cultures."[16]

Achieving cultural expertise and gaining a thorough knowledge of the populace can be a daunting task. Culture encompasses all aspects of society, and determining the training and resources for the constituent areas—geography, language, customs, religion, law, and folklore—can be extremely difficult, especially since a tactical misstep will have strategic implications. While the idea that "that every Marine is a collector" might hold true, the Marine must be properly trained in order to recognize the nuances of the local population response when out in the field. Patrolling and observing the local inhabitant's normal daily activities provides more information than one would expect. Appendix D provides two instances where cultural information was applied directly to operations. They are essential for a non-kinetic approach, an ideal solution to defeat non-Western opponents "who are transnational in scope, non-hierarchical in structure, clandestine in their approach, and operate outside the context of nation-states."[17]

A Non-Kinetic Approach

The role of the cultural anthropologist as a military advisor can be complex, as he or she is charged with studying culture in "meaningful, organized ways, by observing detail, by participating in the culture at hand, by comparing their findings to other cultures, and by using theory to comprehend in its fullest a particular worldview, system of behavior, belief, and psychology."[18] In doing so, the cultural anthropologist must avoid the tendency of social scientists to make generalizations, ignoring the unique aspects of a specific culture.

Current efforts within DoD to strengthen the cultural approach include service academy funding for students to attend cultural immersion programs abroad, teaching facilities upgrades, language and culture faculty expansion, cultural anthropology integration within the Pentagon's Office of Net Assessment, and development of computer and Web-based training that promote cultural awareness training through interactive courseware. In addition, DoD has created the Human Terrain System, which is aimed at addressing "cultural awareness shortcomings at the operational and tactical levels" and dealing with the "social, ethnographic, cultural, economic, and political elements of the people among whom a force is operating."[19] This is a key step in the right direction.

Getting Culture on Board

Since 2003, scholars, politicians, military experts, and ordinary service members have contemplated the most effective way to fight in an IW and counterinsurgency conflict. As Sun Tzu wrote, "If ignorant both of your enemy and of yourself, you are certain in every battle to be in peril."[20] Applied to the military needs of the United States today, this dictum highlights the importance of understanding the culture of the people with whom we are fighting, developing, and/or operating. Colonel Christopher C. Conlin, commanding officer of 1st Battalion, 7th Marines, led his battalion into Baghdad and was expected to conduct transition operations within his area of operation in 2003. In Colonel Conlin's lessons learned report, he states:

> Read more than field manuals. Understand the local culture, political history, and the basics to managing a successful government. Do cultural studies like *The Arab Mind* (Raphael Patai) and the Koran are just as critical as Sun Tzu and Clausewitz in transition operations.[21]

In October 2004, Arthur K. Cebrowski, the director of the Office of Force Transformation, concluded that "knowledge of one's enemy and his culture and society may be more important than knowledge of his order of battle."[22] Why have efforts to obtain detailed cultural knowledge suddenly become so important? One major reason is due to an evolution from conventional warfighting methods of the Cold War to the IW conflicts today in which the combatants live amongst the civilian population. Examples from operations in Iraq, Afghanistan, and Somalia demonstrate that our former "heavy" military doctrine is not effective on a 21st Century IW battlefield.

This trend of the Marine Corps/DoD addressing their need to learn about other cultures is a move in the right direction. There needs to be a push to develop within the military services and DoD many of the requisite skill sets and capabilities cultural anthropologists already possess. Where do we begin with the integration of DoD and the world of cultural anthropology? A step toward integrating DoD doctrine with cultural anthropology may be seen in the joint-sponsored conference on IW and counterinsurgency held in September 2006 by the Department of State (DoS) and DoD. The purpose of the conference was to discuss best practices from a wide array of political, social, military, and economic subjects to help develop a new counterinsurgency framework. The conference had experts from the United States, United Kingdom, and Australia. One expert in cultural anthropology provided the following observations and recommendations: "Comprehensive knowledge of the human terrain; DoD has not funded this since 1965; a database of socio-cultural knowledge; create an organization dedicated to production and dissemination of knowledge of foreign societies to serve our government's counterinsurgency and related missions."[23] These recommendations support developing institutional, academic, and training programs at the tactical, operational, and strategic levels for DoD and supporting agencies.

The Reinvention of the Wheel:
Culture in Past Military Operations

From the Revolutionary War to the present day, American military services have frequently fought in some form of IW conflict.

In each one of these conflicts, the services were slow to recognize the importance of understanding the culture. The Banana Wars in the early 1900s, in which the United States conducted a series of occupations, police actions, and interventions in Central America and the Caribbean, were notable exceptions. The United States owes its successes in these wars to the services' learning and understanding the culture of people that they were fighting and protecting. The Marine Corps *Small Wars Manual* is the direct product of the lessons learned from these wars. It dedicates an entire section to civil relationships, which discusses the importance of understanding the cultural environment, people, and economics of the country.[24] Yet, the military services and DoD exhibit a certain pattern of behavior in such conflicts: the services develop cultural education training programs and doctrine that have immediate relevance but drop them and shift focus as soon as they perceive the operational need for them is over. Unless the United States plans on fighting wars on its own soil—something that has not happened since the War of 1812—it will always need the ability and infrastructure to understand foreign cultures since at the present and near future, the United States military performs almost exclusively an expeditionary role.

Fortunately, there seems to be some recent headway against this trend. Owing to recent conflicts in Iraq and Afghanistan, the Commandant of the Marine Corps established the Strategic Vision Group (SVG) in 2007 to assist the Marine Corps to posture itself for the future.[25] The SVG developed the *Marine Corps Vision & Strategy 2025* report; in it one of the stated goals was to develop and perpetuate training methods to understand and defeat adversaries in complex conflicts:

> We will go to greater lengths to understand our enemies and the range of cultural, societal, and political factors...Our training and education programs will provide skills that enable civil-military and combat operations...particularly important in complex environments...the ability to conduct both types of operations, simultaneously, is the essence of the force as a "two-fisted fighter"—capable of offering an open hand to people in need...in an irregular warfare environment.[26]

The Marine Corps recognizes that the future will present similar cultural issues in both kinetic and non-kinetic operations. Should the military services fail to address the issue of developing courses of study and training on the uses of cultural anthropology, they may find more instances of approaching the local populace in antagonistic way, as a Marine Corps officer discussing his recent tour in Iraq remarked:

> My Marines were almost wholly uninterested in interacting with the local population.... We relieved the soldiers from the 82nd Airborne Division, and their assessment was that every local [national] was participating or complicit with the enemy. This view was quickly adopted by my unit and framed all of our actions.[27]

Such a broad, poorly-informed stereotyping of the local populace is very common among our servicemen and women when they are not prepared to understand the culture in which they are operating. Once these mistaken attitudes are established, they can easily perpetuate from one unit to another during relief in place (RIP) transitions. When military units are ignorant of the social environment they are operating in, they are essentially fighting with one hand tied behind their backs, where even the most well-intentioned approaches may be rendered ineffective. As one Marine found out: "We were focused on broadcasting media and metrics. But this had no impact because Iraqis spread information through rumor. We should have been visiting their coffee shops."[28]

Educating for the Future

Providing American servicemen with the skills to understand a new culture cannot be done through a single seminar or two-week course; it must be a career-long process. The future leaders on the Joint Chiefs of Staff are already on active duty today as captains and lieutenants; these future leaders must understand the value of cultural anthropological approaches to war, or the United States military will repeat the same hard lessons of the past. There is a need to start educating all military servicemen, both officers and enlisted, on the value of culture in IW for the military of tomorrow.

Service members must do more than simply duplicating the formula or the answer without understanding the operational culture; they must understand the reasoning behind the process. To ensure that this is so, such operationally-relevant cultural training must be taught and reinforced throughout the servicemen's entire careers. Figure 19—developed by the former director of the Training and Education Command, Major General T. S. Jones—has been modified to demonstrate that cultural training, like leadership skills training and unit training, needs to be taught throughout servicemen's military careers to maintain a high proficiency level. The Officer Professional Military Education Policy (OPMEP) states that it is necessary to foster a "learning continuum that ensures our Armed Forces are intrinsically learning organizations" to ensure that officers, "over the breadth of their careers, become the senior leaders."[29] The professional military education wedge in Model 1 illustrates how PME takes the place of skill progression training in a typical military career. This reflects the changing requirements for a serviceman's skill set, evolving from battlefield tactics early in his career to staff planning and leadership positions later on.

The goal of this training and education is to populate the 2025

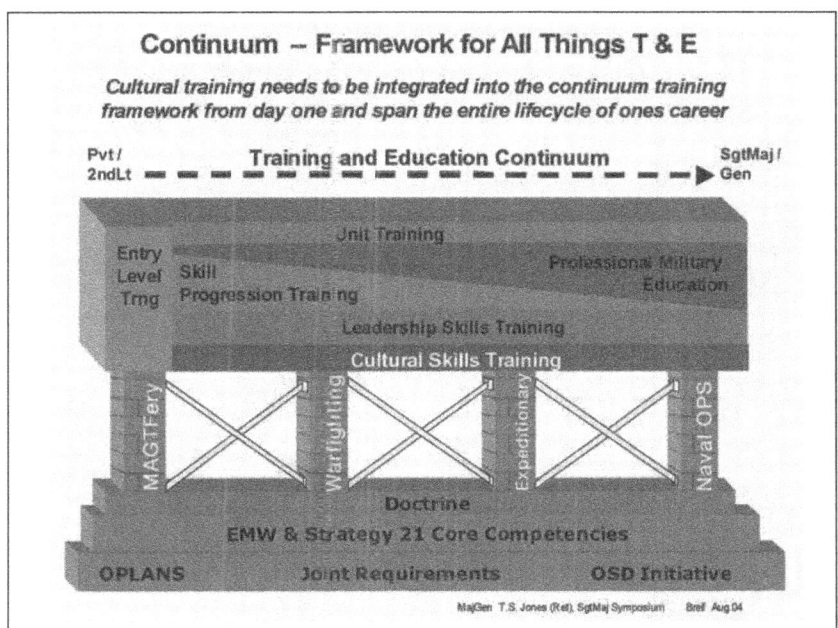

Figure 19

130 *Applications in Operational Culture*

Marine Corps with Marines who are educated in the cultural process in order to meet the demands of operating at all three levels of war—tactical, operational, and strategic. The PME system can be used to implement requirements and instructional guidance for both officers and enlisted servicemen as they progress through their military careers and take on greater responsibilities at the higher levels of war.

The cultural skills training and leadership skill training banners illustrated in Figure 19 reflect the importance of both abilities within Marines career progression. The importance of leadership skills training has always been emphasized as one of the most important skills for any leader and is taught throughout a Marine's career. Cultural training should be treated similarly. To be proficient in recognizing and understanding a culture, Marines must continue to study, develop, and practice anthropological approaches to culture. Therefore, it is important that cultural skills training be part of the PME process to ensure the continuous development throughout a Marine's career.

Figures 20 and 21 reflect the proposed enlisted and officer career progression schedule for operational culture education. Both

Officer Career Progression of Operational Culture PME				
Rank	Course	School	Operational Level	Assignment
Second Lieutenant	Introduction	The Basic School	Tactical	Student
First Lieutenant	Introduction II applied towards MOS	MOS/ Unit	Tactical	Student Platoon Cmdr
Captain	Intermediate	Expeditionary Warfare	Tactical Operational	Company Cmdr B Billet
Major	Intermediate II	Command and Staff SAW	Operational	Joint XO Staff Officer
Lieutenant Colonel	Advanced	SAW MCWAR	Operational Strategic	Battalion Cmdr Joint Staff Officer
Colonel	Master I	MCWAR	Operational Strategic	Regt Cmdr Joint
General Officer	Master II	CAPSTONE	Strategic	Div/Group/MEB/ MEF/ JCS

Figure 20

Enlisted Career Progression of Operational Culture PME				
Rank	Course	School	Operation Level	Assignment
E1-E3	Introduction	Marine Combat Training/ MOS/Unit	Tactical	Fire team Squad
E4-E5 NCO	Introduction II	NCO Academy Unit	Tactical	Squad Ldr B-Billet
E6-E7 SNCO	Intermediate	SNCO Academy Unit	Tactical Operational	Platoon Sgt Company Joint
E8-E9 SSNCO	Advanced	Advanced Schools	Tactical Operational Strategic	Co/Bn/Div/MEF/ Joint

Figure 21

tables provide the appropriate serviceman rank, the level of instruction, which schoolhouse will conduct the education, the corresponding operational level of war, and the career/future assignment of the Marine.

The first levels of instruction for junior officers and enlisted would be introductory courses held at the Basic School for the former, Marine Combat Training facilities for the latter, and MOS schools for both. These courses should address basic concepts of culture, cultural events at the tactical level of war, and the value of information on the ground. The second level of instruction would consist of intermediate courses for captains, majors, and Staff NCOs (E-6 and E-7s). These include instruction for both the tactical and operational levels, between the rank of captain and major, with the latter focusing on the operational level. The third level is for senior field-grade officers and senior staff NCOs, and would contain the Advanced and Master I Courses, which focus on how cultural models can be applied on the operational and strategic levels of war. The Master II-level course is the final course in the PME schedule, which is for general officers and shapes their understanding of how cultural aspects can be shaped on the strategic level.

Don't Forget Interagency

It is also important to note that low-intensity operations are not solely a military responsibility. As Dr. Jeffrey "Jeb" Nadaner, Deputy Assistant Secretary of Defense for Stability Operations, stated:

> Insurgency is here to stay...counterinsurgency is 80 percent non-military (including political, economic and development), and 20 percent military. As a government, the United States needs broad set of tools to fight irregular war. It needs unity of effort across the broad spectrum of government agencies, not just the military.[30]

Like the military services, many of the governmental agencies that partner with DoD need to develop a stronger understanding of the cultural environments for their own operational effectiveness. Since IW conflicts increasingly involve a whole-of-government approach, failure to understand the cultural environment can negatively impact both on an agency's operational plan as well as the military's mission. It is just as imperative that these nonmilitary agencies operating abroad also develop social science training programs and require mandatory attendance prior to deployment. DoD and its governmental agency partners should combine their efforts for developing a national cultural research center and schools to develop and ensure a common curricula used by the military and other governmental agencies. This will strengthen the interagency partnership in future low-intensity conflict and will determine success or failure.

Conclusion

An examination of key elements of IW military operations illustrates why an in-depth understanding of the populace is essential to success and anthropology is the discipline most suited to help the United States Marine Corps understand this populace. The use of cultural anthropology in military matters has a strong history, both in the colonial and military exploits of the British, French, and the United States in past centuries. As such, adopting the cultural approach as a tool is useful—perhaps even indispensable—for the successful conduct of IW operations. If elements within the United States' national security establishment, such as DoD, the military services, and associated governmental agencies, accept this premise and adopt modifications to their training and planning for IW operations, current and future conduct of IW conflicts will benefit from having better information, guidance, and preparation in pursuit of the objective.

Notes

[1] Department of Defense Irregular Warfare Joint Operating Concept, Version 1.0, 11 September 2007.

[2] David Galula, *Counterinsurgency Warfare: Theory and Practice* (Westport, CT: Praeger Security International, 1964), 4.

[3] Ibid.

[4] Ibid., 106.

[5] Ibid., 52.

[6] Robert Thompson, *Defeating Communist Insurgency: The Lessons of Malaya and Vietnam* (London: Chatto and Windus, 1966), 51.

[7] Julian Paget, *Counter-Insurgency Operations: Techniques of Guerrilla Warfare* (New York: Walker, 1967), 176.

[8] Ibid., 162.

[9] U.S. Army Field Manual 3-24, *Counterinsurgency* (Washington, DC: U.S. Government Printing Office, 2006), sec. 3-77.

[10] Gary Ferraro, *Cultural Anthropology: An Applied Perspective*, 5th ed. (Belmont, CA: Wadsworth Thomson Learning, 2004).

[11] Clifford Geertz, *The Interpretation of Cultures: Selected Essays* (New York: Basic Books, 1973), 4.

[12] Ibid., 53.

[13] Robert H. Scales Jr., "Culture-Centric Warfare," *Proceedings*, October 2004, 32.

[14] Ibid., 33

[15] LtCol William D. Wunderle (USA), *Through the Lens of Cultural Awareness: A Primer for US Armed Forces Deploying to Arab and Middle Eastern Countries* (Fort Leavenworth, KS: Combat Studies Institute Press, 2006).

[16] Isaac N. Skelton, Whispers of Warriors: Essays on the New Joint Era (Washington, DC: National Defense University Press, 2004), 133.

[17] Montgomery McFate, Adversary Cultural Knowledge and National Security Conference, November 2004, quoted in the Office of Naval Research *Originator* (http://fellowships.aaas.org/PDFs/2004_1210_ORIGConf.pdf).

[18] Pamela R. Frese and Margaret C. Harrell, eds., *Anthropology and the United States Military: Coming of Age in the Twenty-first Century* (New York: Palgrave Macmillan, 2003), 136.

[19] Jacob Kipp, Lester Grau, Karl Prinslow, and Don Smith, "The Human Terrain System: A CORDS for the 21st Century," *Military Review*, September-October 2006, 9.

[20] Sun Tzu, *The Art of War* (New York: Penguin, 2005).

[21] Col Christopher Conlin, "What Do You Do for an Encore?" *Marine Corps Gazette*, September 2004, 80.

[22] Montgomery McFate, "Anthropology and Counterinsurgency: The Strange Story of their Curious Relationship," *Military Review*, March-April 2005, 2, 24.

[23] Montgomery McFate, Counterinsurgency in the 21st Century: Creating a National Framework Conference, September 2006.

[24] U.S. Marine Corps, *Small Wars Manual* (Washington, D.C.: U.S. Marine Corps, 1940), NAVMC 2890, ch. 1; par 1-28-29, pp. 41-44; ch 5; par 5-13-25, pp. 13-20.

[25] *Marine Corps Vision & Strategy 2025* (Washington, DC: U.S. Marine Corps, 2008), 23.

[26] Ibid., 25.

[27] Montgomery McFate, "Can Social Scientists Redefine the 'War on Terror'?" *New Yorker*, 16 December 2006.

[28] Ibid.

[29] Chairman of the Joint Chiefs of Staff Instruction, "Officer Professional Military Education Policy" (Defense Printing, 7 August 2007). The OPMEP applies to the Joint Staff, the National Defense University (NDU), and the military services.

[30] Dr. Jeb Nadaner, Counterinsurgency in the 21st Century: Creating a National Framework Conference, September 2006.

Chapter 6

Operational Culture: Is the Australian Army Driving the Train or Left Standing at the Station?

Lieutenant Colonel Steven Brain, Australian Army

Lieutenant Colonel Steven Brain entered the Royal Military College, Duntroon, in 1993 after completing his bachelor of arts and graduate diploma of defence studies. In his first year as a lieutenant, He was a platoon commander in Company A of the 2d/4th Battalion in Rwanda as a part of the United Nations Assistance Mission (UNAMIR II). As a captain, he served in a variety of regimental appointments, including the adjutant of Monash University Regiment and the 1st Battalion. He also served as an instructor at the Royal Military College.

As a major, Brian took command of Company A, 1st Battalion, Royal Australian Regiment. It was in this capacity that he deployed to East Timor as part of Operation Citadel. During this time, his company was deployed on the border with Indonesia and was given a security mission. On completion of this posting, Brain deployed on Operation Catalyst-Iraq as an Australian embedded officer. He worked as a battle major (current ops) in the Strategic Operations Centre on the Multi National Force-Iraq Headquarters. In 2006, he came to the United States to attend the U.S. Marine Corps Command and Staff College. He was subsequently selected to study at the U.S. Marine Corps School of Advanced Warfighting. After graduation, Brian was promoted to lieutenant colonel and is currently the staff officer grade one of the Enhanced Land Force Capability Implementation Team within the Australian Army Headquarters. He has recently been selected as the commanding officer 2010 designate of the 51st Battalion, The Far North Queensland Regiment.

> *The outcome of future conflict will increasingly be decided in the minds of...populations rather than on the battlefield. Therefore, combat operations can no longer be seen as the decisive phase of conflict.*
>
> Australian Army, *Adaptive Campaigning,* November 2006

Twenty-First Century military experience has been characterized by many factors: maintenance of combat tempo, adaptation to complex operating systems, and large-scale coalition operations. With warfare currently consisting of a mixture of conventional and unconventional conflict, it is safe to assume that future conflict will also consist of the same mix—at least to some degree. One factor that is certain is that all armies of the United States, Britain, Canada, and Australia (ABCA)[1] consider culture to be vital across the spectrum of operations.

Despite the importance of culture initiatives in many Western militaries, in the Australian Army today, no cultural training is currently formalized. Individuals or units notified for deployment have to invent their own training objectives; this lack of institutional training necessitates adaptation in the field or while on operations. The decisions regarding training objectives, how much training to conduct, and how to adapt while on operations is largely left to the commander's discretion.

The Australian soldier has long relied on the "good bloke" factor when dealing with foreign cultures. While this approach has worked until the present time, as the world changes, the "good bloke" factor is no longer enough. If the Army wants to improve the way it operates and ensure it is effectively prepared for future warfare, the Australian Army must change its methods. In order to accomplish this, the Australian Army needs to define cultural capabilities and train individuals and units to be able to fulfill that capability

The Chief of the Australian Army, Lieutenant General Peter F. Leahy (Service Chief equivalent in the U.S.), recognized this need for change in his "Intent," which he publishes each year. The 2007 version of this intent included new direction on cultural awareness. The three key strategic drivers that Lieutenant General Leahy is concerned with presently are: 1) societal/human issues; 2) ongoing globalization issues; and 3) environmental issues. This chapter discusses the societal/human issues.

The Chief of the Australian Army states that there is an enduring aspect to cultural perspectives of warfare and this should be weighed against the transient nature of technology. This statement highlights the importance the Australian Army is placing on operational culture. This chapter proposes a method for implementing the Chief of the Australian Army's directives for culture: outlining both changes in training and in military structures to accomplish the Chief's objectives.

Defining Culture

In anthropological terms culture is the sum total of ways of living built up by a group of human beings that is transmitted from one generation to another. Militarily, however, this definition is insufficient and does not really assist the Australian warfighter because it does not describe reality as it will exist on an operational deployment. For the warfighter, culture can be defined as "a shared world view and social structures of a group of people who influence a person's and a group's actions and choices."[2] These actions give a warfighter something observable on the ground that can be analyzed and incorporated into operational planning. The warfighter is not going to be concerned with all aspects of culture, but only those aspects that influence the area where warfighter operates. Therefore, operational culture as a term is of greater use to the warfighter than simply culture: that is, "those aspects that can influence the outcome of a military operation or, conversely, those military actions that influence the cultural balance within an area of operations."[3]

Other key definitions that need to be addressed are: human terrain, social environment, and cultural factors. Human terrain encapsulates "those cultural aspects of the military environment, which, due to their static nature, can be visually represented on a geographic map. Human terrain is static with respect to change over time and rigid with respect to fluid human relationships."[4] Once the human terrain has been mapped, the social environment can be manipulated (or in military terminology: shaped), including the human relationships and interactions among the people.[5] Cultural factors are aspects of society that have the capacity to affect military operations. They include: religion, ethnicity, language, customs, values, practices, perceptions and assumptions, power and influence sources, and driving causes like government, political and social grouping structures, economy, and security. All these factors affect the thinking and motivation of the individual or group and make up the cultural terrain of the battlespace. However, it is important to note that not all factors are applicable to all operations, and additional factors may need to be considered as necessary.[6]

Levels of Expertise.

The study of operational culture is popular and gaining mo-

Figure 22

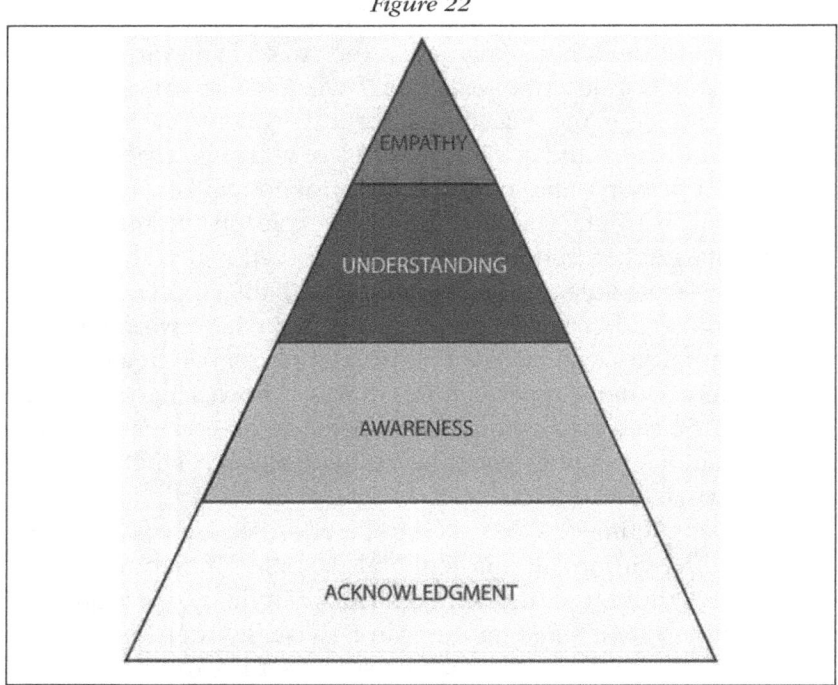

Levels of Operational Culture
This figure depicts the hierarchical relationship within operational culture. It also shows the size of the Australian Army expected to employ that level (in relative terms).

mentum in most Western militaries at the moment, but much of the terminology across service or coalition boundaries is being used in a contradictory manner. This chapter will offer clear definitions and also propose a model demonstrating that a hierarchy exists in the application of these terms within the operational culture construct. As depicted in Figure 22, key operational culture terms include cultural acknowledgment, cultural awareness, cultural understanding, and cultural empathy.

Cultural Acknowledgment is the basic acceptance of the direct importance of cultural operations. Cultural acknowledgment means admitting that cultural factors will influence the future battlefield and, by working to improve and then to exploit those skills in the soldier, the Australian Army will have a better chance of operational success. This acknowledgement illustrates various cultures are different and we should strive to learn more about them. This level is the minimum level of cultural understanding the Australian Army should expect from junior ranks during peacetime training. By achieving cultural acknowledgement, soldiers are prepared for higher level learning and are provided a foundation for developing professional and personal interest in cultural education.

Cultural Awareness is the knowledge of cultural factors and a comprehension of their impact on the planning and conduct of military operations. Cultural awareness results from both standardized and specific training.[7] The 2006 ABCA Standardization publication on *Cultural Awareness* claims that cultural awareness can reduce battlefield friction and the fog of war, thus improving a unit's mission accomplishment. Awareness gives insight into the intent of various actors within our battlespace and the way these actors interact within groups. The Australian Army can use this information in its attempt to shape the battlefield towards more favorable conditions or build rapport to prevent misunderstandings that might prevent us from accomplishing our mission.[8] Such information will also support planners in developing centers of gravity to ascertain critical vulnerabilities and will assist in campaign planning and the proper allocation of resources. Cultural awareness is the minimal level of cultural knowledge we should expect from soldiers on an operational deployment (see Figure 25).

The Deputy Chief of the Army, Major General John P. Cantwell, in his "Planning Guidance for Development of Cultural Understanding Capability in the Australian Army," dated November 2007, has defined **Cultural Understanding** as the capacity for active study of human and cultural influences affecting all decision-making and actions in the operating environment, in order to optimize one's own decision superiority through empathy.[9] This definition refers to a deeper awareness of the specific culture that allows general insight into the thought processes, motivating factors, and other issues that may be scrutinized for planning purposes.[10]

Dr. Patrick Guinness, head of the School of Archaeology and Anthropology at the Australian National University, stated on 16 August 2007, during a trip to Marine Corps Base Quantico, Virginia, that it is crucial to understand the cultural environment in which our Army must operate. We must go beyond the understanding of the societal and cultural environment and indeed "empathize," that is, identify mentally with and so completely comprehend a culture.[11] The Chief of the Australian Army has put emphasis on empathy in his "Commander's Intent" to the Army. In a direction that is similar to Dr. Guinness' premise, the chief urges commanders to strive for cultural empathy in training and on operations.

Cultural Hierarchy

The cultural knowledge hierarchy depicts what you can achieve the more time is allocated toward it.

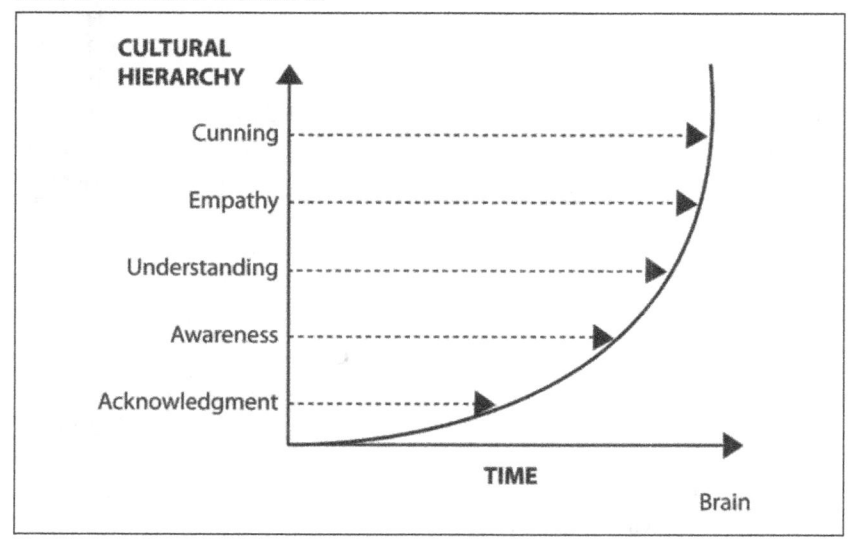

Figure 23

While the three previous terms are currently in use in the Australian Army, the author proposes one final important concept: cunning. **Cultural Cunning** is the real crux of the warfighter's capability. The application of cultural cunning occurs when key personnel are able to integrate information gained into understanding the cultural factors. This fusion of information and military requirements will support mission success with the application of cultural cunning. It is well and good to have operational situational awareness and to culturally understand the factors in your operating environment. However, having the astuteness to use your knowledge correctly and ascertain outcomes that are advantageous to the warfighter should be the goal of key personnel (commanders and planners). This is what they should be striving for in the hierarchy of operational culture. In Figure 23, cultural cunning would sit on top of the pyramid, depending on all the factors that precede it, in order to be employed.

The cultural knowledge hierarchy as depicted in Figure 23 portrays a curve that increases in knowledge as time moves forward. The more time one has to study a culture, the higher he should be able to reach in the hierarchy.

Culture and Future Warfare

Challenges

It is difficult to predict what future warfare will look like. Failure to predict well explains why so many militaries in history have been branded as "prepared to fight the last war" when a new type of conflict emerges. The nature of war will remain enduring; however, the characteristics will change. Regardless of technical innovation and natural emergence, recent trends suggest conflict in the future will increasingly involve diverse actors, all competing for the allegiance of targeted populations. This is not drastically different from contemporary warfare; it does, however, suggest that this form of warfare is here to stay. Montgomery McFate describes the adversary of future conflict as "non-Western in orientation, transnational in scope, non-hierarchical in structure, and clandestine in approach; and it operates outside of the context of the nation-state."[12] This description of future warfare is what many have come to define as "irregular warfare."

A common argument being debated in professional military journals presently is the degree to which modern militaries should be preparing forces for irregular warfare as opposed to conventional high-end operations. With the operational tempo so high at the moment, and with no foreseeable end, preparing for one is usually at the expense of the other. There is a high probability that they will co-exist, but the degree to which one is more important than the other is unknown. The author agrees with the U.S. Department of Defense Joint Operating Concept for Irregular Warfare (2007), which claims it will be a supported/supporting relationship.[13] In short, both aspects of conflict will be present: either conventional conflict will support the irregular warfare piece, or the irregular operations will support a conventional fight. As such, fighting on the future battlefield will depend not just on traditional military process, but also on understanding social dynamics, such as tribal politics, social networks, religion, and cultural norms. The key to favorably shaping the battlefield and striving for victory will be in the minds of the populations rather than pure kinetics. As the Australian Future Land Operational Concept, *Adaptive Campaigning*, states: "Combat operations can no longer be seen as the decisive phase of the conflict, and as a result, a comprehensive approach to future land force operations is required."[14] The force of the future will require soldiers who are patient, persistent, and culturally astute to be able to influence the operating environment in ways that increase the Australian Army's chances for battlefield success. This operating environment can be viewed as a complex adaptive system.

Complexity

A precise definition of complex adaptive systems can be found in Harold Morowitz and Jerome L. Singer's publication *The Mind, The Brain, and Complex Adaptive Systems*. Their explanation of complex adaptive systems describes the system as involving numerous active actors, existing in many hierarchical layers, whose collective behaviors shape the whole.[15] Such aggregate behavior is non-linear, hence it cannot be simply examined from summation of the individual component parts. Each of these actors is individually diverse; and, when applied to the military context, it means each influence or flux introduced onto the system is uniquely different, each with its own unique impetus.[16]

An important feature of complex adaptive systems is the sensitivity to even small perturbations or events. One single impetus, either an input or extraction of an agent or energy, can produce a very broad range of reactions or responses. If the same impetus is repeated in the exact same measure, one still cannot guarantee the same response. This makes it exceedingly difficult to predict a response of the system or even to establish likely scenarios.[17]

This has significance to military commanders or planners as they attempt to shape an outcome by military action. Each of the cultural factors listed earlier is related to the other factors in a complex way. Soldiers must understand the possible adaptation outcomes they are likely to encounter. Regrettably, understanding this entire system is impossible; the best military commanders can hope for is to understand where in this system they can input energy or influence to have an effect and analyze what is believed to be the likely result as the system adapts. An example of this in Iraq is understanding the Arab male requirement to save face and not be seen as abetting the coalition. As such, anonymous phone call centers have achieved a degree of success for the coalition. To be able to tell where a commander can have the most influence will require a certain degree of cultural empathy, then the commander can apply cultural cunning towards the actors and their cultural terrain.

The Australian Army: Operational Culture in Education and Training

The first question that must be asked is: Does the Australian Army's conventional approach to training directly transfer across to fighting within the adaptive system of irregular warfare? The direct and simplest answer is "yes." The Australian Army has the skills, knowledge, and attitude to be able to operate successfully in this environment. However, the question remains: Can we do it better?

In order to accomplish this, the Australian Chief is calling his target outcome "understanding cultural empathy." He has decreed a three-tiered approach: first, generic inculcation of cultural empathy to occur at individual and unit collective training levels; second, focused mission rehearsal exercises; and third, specific in-country cultural advisor assistance for the deployed units.

Education and Individual Training

There are two approaches the Australian Army can take to accomplish cultural training: Improve the cultural awareness of every single Australian Army soldier; or, develop a pool of specialists. Generic inculcation should begin at the individual's military basic schools and should be enhanced through courses—either specialist courses or promotional courses—throughout a soldier's career. Until now this has been an ad hoc measure, implemented by astute staff recognizing a training requirement. To formalize training the all-corps soldier, the training continuum needs to be amended and improved to account for culture. These continua begin at basic training and continue through a soldier or officer's professional military education (both mandatory career courses and mandatory promotional courses). Within the next year, it is recommended that a "training needs analysis" be conducted to determine the knowledge, skills, and attitudes an Australian soldier needs to be able to execute operations in complex human terrain.

To influence the future battlefield, cultural skills should be defined, as they are not the traditional or core areas of expertise for most military forces. Once defined and ratified, specific skill sets should be implemented as training objectives into the soldier and officer all-corps training continuum to achieve the required results.

Peacetime Requirements

Figure 24 depicts, in a standard peacetime training continuum, that a soldier should be trained to achieve cultural acknowledgment. Officers and senior noncommissioned officers should be trained to achieve cultural awareness. Key personnel in units and higher, such as commanding generals; commanding officers; operations officers; future planners; and, civil affairs teams to name a few, should be trained to achieve cultural understanding.

To achieve the prescribed levels in Figure 24, key personnel

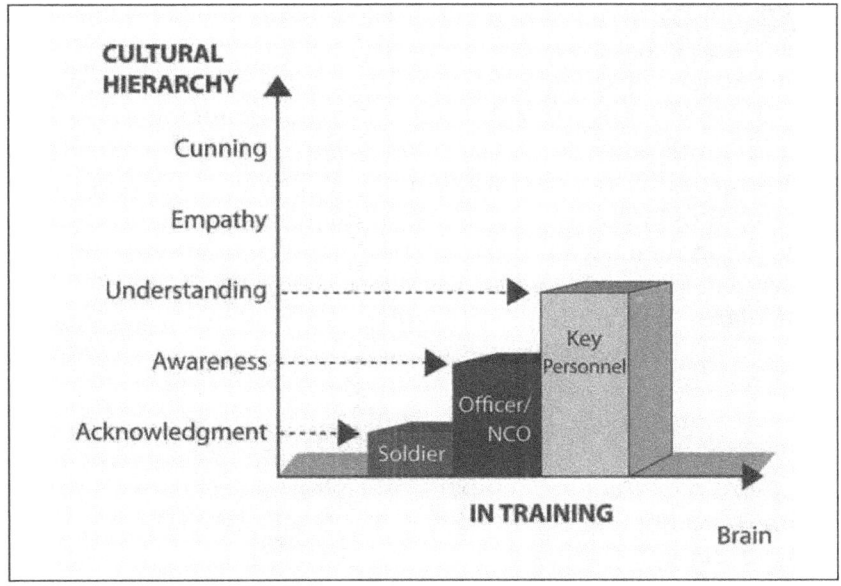

Figure 24
Standard Peacetime Training Objectives Applied Against the Cultural Hierarchy.

need to be briefed by higher headquarters on likely location scenarios for possible deployment. Once deployment has been forecast as likely or imminent, it narrows and focuses the knowledge service personnel are required to have. Soldiers should be trained to the cultural awareness level necessary to operate in the specific cultural operating environment of the country (or region) of the deployment. Key personnel mentioned above now need to develop cultural empathy to completely synchronize their attitudes with the culture in which they will find themselves and their units operating. Once this is achieved, cultural cunning can be employed to best shape the entire battlefield to advantage, albeit with kinetic or nonkinetic capabilities. These capabilities may or may not include language training.

It is not easy to say the Australian Army should focus on any one country and develop a cultural awareness program or language improvement program for a specific set of cultures alone. To compensate, Australian Army units will run short courses to improve soldiers' individual language skills. These could last anywhere from two days to two weeks. Soldiers who have an identified talent for a particular language may be sent to the Australian Defence Force (ADF) School of Languages (LANGS). At this school, they would participate in a three-month language program aiming to develop their linguistic skills as well as cultural empathy. The course would be instructed by people of an ethnic background relevant to the country they are studying. These soldiers would then return to units to run their own language training and cultural presentations. Individual soldiers identified to deploy on a detailed mission to a particular region could then be subjected to a year-long "cultural immersion" course at LANGS to prepare for a job in a particular country.

These language training opportunities are extremely scarce and may affect one or two soldiers per major unit in peace-time and a few more preparing for deployment. Unfortunately, a three-month course does not train an individual adequately for a deployment as an interpreter.

There also exists a different kind of individual training regimen. An individual "augmentee" may be ordered into a specific theater to supplement a unit or indeed represent Australia in a Coalition billet (as the author did as a Battle Major in the Multi-National Force Iraq Strategic Operations Center in 2005-2006). This individual will have to complete organized pre-deployment training with the 39th Personnel Support Battalion, Force Preparation Company. For the author, this included area and situation briefs, Islamic cultural briefs, and operating forces briefs. This preparation is enhanced by in-country "Reception, Staging, On-forwarding and Integration" (RSO&I) training. Currently, this training lasts three days and includes one "double lesson" on regionally specific cultural information. To maintain a credible RSO&I training program, the Australian Army could use the assistance of cultural anthropologists. The lessons presented were given (in the author's case) by an Australian Iraqi teaching language skills at LANGS. Although her language skills and cultural knowledge were genuine, her ability to be able to pass relevant information was marginal. However, a cultural anthropologist would be able to break down the physical and moral nodes that exist for the culture, place them in context with the region individuals are deploying to, and synthesize the information into something useful for soldiers of all ranks.

Education and Unit Training

The second part of a soldier's cultural education will occur at the unit level. This will come in the form of selected countries or areas to which a unit is likely to deploy. For an Australian soldier, this will often be the immediate "inner arc" for strategic national interests; it extends through the Indonesian Archipelago, East Timor, through Papua New Guinea to the islands of the South West Pacific. Any sustained conventional attack on Australia would need to come from or through these islands. Therefore, it is the region that the Australian Army trains to operate in during routine training cycles. Australia has deployed troops most frequently to these areas in the last ten years—not to conventional conflicts, but in a security capacity: quelling coups, countering insurgency, or combating militias. Such deployments have included Cambodia, Papua New Guinea (Bougainville), the Solomon Islands, and East Timor.

Individual training will differ from collective unit training. Within a unit, the commanding officer will give guidance on cultural train-

A Standard Operational Deployment Objectives Applied Against the Cultural Hierarchy.

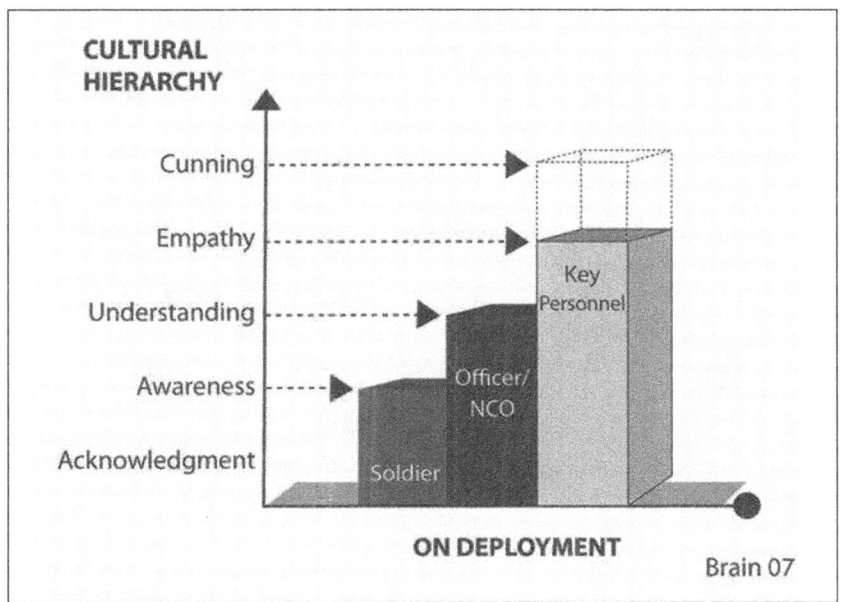

Figure 25

ing within his commander's directive issued at the start of each training year. If a particular unit has been designated to deploy then the training program will be tailored toward that country. Currently, units have posted a soldier with previous experience (more often than not with either military or civilian encounters) in the culture where the unit is likely to deploy. Therefore, there are a number of soldiers within units and brigades who have experiential knowledge to pass on regarding operational culture. This can be formalized in mounting directives or simply organized by the unit prior to deployment. This training is not mandated and is purely at the commanding officer's discretion. The Army headquarters should direct the exact cultural training a unit needs to complete before deployment. Currently, the only direction is concerned with the mission rehearsal exercise (MRE).

The MRE is usually the last in several months' worth of unit preparation. It should occur in the middle third of the allocated preparation training time for a unit to be able to fix the errors identified before the unit deploys. This change in organizational attitude needs to occur at the operational level in order to affect the tactical level. To be suitably prepared for future warfare, units need to meet specific criteria in cultural education and training. One hundred percent of deploying soldiers need to be competent in cultural objectives specific to their position and rank in the soldier all-corps training continuum. This should be completed prior to being allowed to partake in the MRE. The Operations Branch of Army Headquarters could issue a directive stating the collective cultural training that must be complete prior to deployment. This would ensure an Army-wide standard to meet operational level intermediate goals. These goals can be re-examined periodically and modified to focus on the cultural areas where units are deploying, for example the current Operation Catalyst (U.S. Operation Iraqi Freedom). This would be the prerequisite for collective unit training to be tested at the MRE.

The most significant training a unit conducts in a formal setting and is facilitated by the Combat Training Centre (CTC) is the MRE. The CTC attempts to replicate the theater into which the exercising unit is about to deploy. It aims to expose soldiers to the towns, the culture, and to the worst case threat that could present itself at any stage. The cultural aspects are highlighted by several methods.

First, role players replicate the major players a unit must deal with in theater. These include the national army, police, tribal and religious leaders, local and provincial government leaders, threat forces, as well as the general community. There is the drawback that the role players are not ethnic locals, such as Iraqis or Afghans, but soldiers being employed in that role. As such, only English is used throughout the MRE. This is recognized as a serious limitation in the MRE function, but is one that is extremely difficult to overcome considering the small numbers of linguists currently in the Australian Army and the high cost of employing civilians suitable for the role. The only time this varies is when the CTC is able to source a role player from LANGS who can converse in the language applicable, otherwise it is limited.

With some direction of resources, the CTC could improve the soldiers' cultural awareness, an officer's understanding, and practice the unit's key personnel in application of cultural cunning. These resources would allow a full brigade to engage in a mixture of live and simulated events. This would allow a larger number of troops to be exercised rather than the one battalion that is currently rotated through the MRE. With such increased numbers, linguists could then be used as both exercise role players and as interpreters for the officers and NCOs on a regular basis and to add credibility. This would enable on-the-spot education and reinforcement of language skills. The CTC would need to be capable of this expansion by 2012 to prepare troops for the future of warfare in 2015.

Deployed Forces

In theater, units will conduct ongoing training, such as after action reviews (AARs), to determine the causes and effects of an action and what needs to be improved. This is as applicable to high-end combat as it is to engaging the civil community in a local market. Further training in languages and any specific cultural issue that has caused concern would be presented throughout the unit. Continuation training in-theatre is also a key to maintaining current information across the unit. As formal cultural advisors are placed into unit headquarters in the future, this will greatly enhance ongoing theater training.

Australian units will use their various headquarters staff cells in

150   *Applying Operational Culture*

a manner similar to other ABCA armies and produce various forms of stakeholder analysis for the commanding officer and his command team. This tries to depict local linkages between factions and interested parties. As knowledge grows from meetings, patrols, or conversations, it is amended and updated. In this way, a unit can leverage its knowledge of the local community and local nuances to the unit's advantge. Often units will not see positive results for long periods of time. This can be extremely frustrating, especially as there are no metrics to assess the productivity of the unit's actions.

Structural Change

Organizational change is recommended in the form of establishing the human systems company (HSC). This company should be raised as another company of the 1st Intelligence Battalion, a unit working directly for the 1st AUS Division. In a perfect world, these organizational positions would all be intelligence corps personnel, but this capability would be too hard to achieve in both force development and career management, even given a five-year lead time. Therefore, as a compromise, the HSC would consist of all corps officers and NCOs. Force structure permitting, the HSC would be commanded by a lieutenant colonel post-command. The cell commander would be a major and the detachment commander a captain (with his second-in-command a senior warrant officer to facilitate half-team deployments).

The aim of the unit is to have five "cells" focusing on a region of the world (see Figure 26). Each region is broken down further into "detachments" that focus on a particular culture or country. The detachments would consist of six-man teams that can deploy to an operation with the task force or battle group headquarters to provide advice to headquarters planners (especially the commanding officer).[18] There would be two deployable teams within the six man-detachment to provide some immediate redundancy. The fifth cell would not be tasked with any region, but would pick up unexpected tasks or assist in drawn out deployments, as in the case of Iraq or East Timor. Once ongoing commitment has been identified, the detachments should have 12 months minimum to focus their training and education on a region.

It is envisioned that the HSC would maintain close analytical

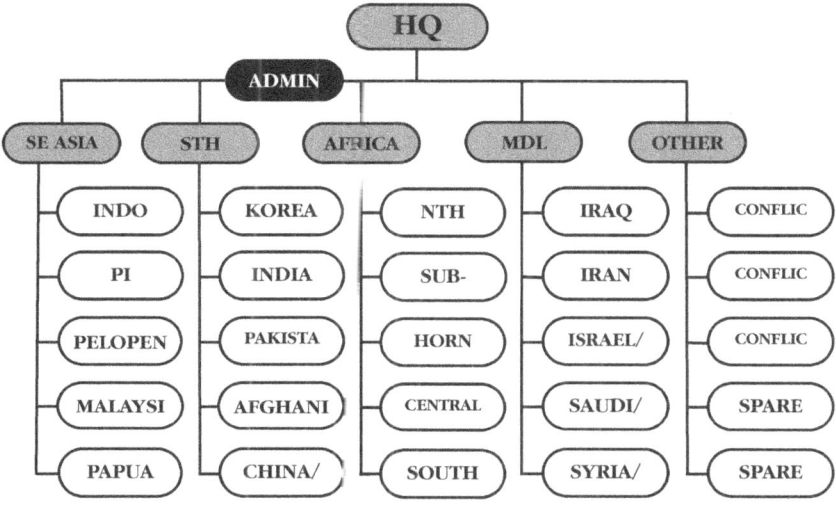

Figure 26
The Human Systems Company

links with civilian academics such as PhD-level anthropologists, social scientists, political analysts, and country specialists. The role of the detachments deployed would be to immerse themselves in the local culture and try to gain cultural empathy so as to assist the unit in using cultural cunning to win the war. This is a crucial reason why military members, not civilians, must work and deploy in the HSC. They understand the military environment and planning requirements but do not struggle with the ethical conflicts facing civilains when information they have provided may or may not be used toward a violent outcome.

Conclusion

Culturally, the success that Australian soldiers have experienced is only due in part to limited cultural training. The natural curiosity of the Australian soldier means that he has a strong desire to engage with "locals" at every opportunity. He sees locals as a source of fascination rather than anger, repulsion, or disdain. This nature is derived partly from the way in which Australians exercise and train. Nevertheless, the Australian Army must accept greater responsibility for individual soldier and officer training and education. Deciding which particular ranks require what level of cultural knowledge is the crucial first step. Carrying this into unit collective training is the second. A third approach is to undertake an organizational adjustment as adding a human systems company into the intelligence battalion. The members of this HSC would become military experts in particular cultures who deploy with tactical level headquarters, allowing commanders and planners to employ cultural cunning to suit the complex cultural terrain in which the Australian Army will operate in the future.

Notes

[1] ABCA: With the ratification of the Basic Standardization Agreement 1964 (BSA 64) on 10 October 1964 by the Armies of the United States (US), United Kingdom (UK), Canada (CA) and Australia (AS), the current ABCA Armies' Program was formally established as the American, British, Canadian, Australian Armies' Standardization Program. In 2004, the US Army signed an MOU with the US Marine Corps that formalized their increasing participation in the Program. As a result, the US is currently represented by a single national position, typically through the senior US Army representative. New Zealand was officially accepted as a full member in March 2006 but the title remained unchanged as the ABCA Armies' Program. The ABCA Program is a vibrant, proactive and evolving organization that reflects and pursues the shared national values and defense goals of its member countries. Today, the focus of the Program is on interoperability, defined as "the ability of Alliance Forces, and when appropriate, forces of Partner and other Nations, to train, exercise and operate effectively together in the execution of assigned missions and tasks." See http://www.abca-armies.org/History, accessed 28 August 2007.

[2] Barak A. Salmoni and Paula Holmes-Eber, *Operational Culture for the Warfighter* (Quantico, VA: Marine Corps University Press), 29.

[3] Ibid., 46, 47.

[4] Ibid., 46.

[5] Ibid.

[6] ABCA Standard 2066, "Cultural Awareness Tactics, Techniques and Procedures," authorized by Col D. A. MacLean, COS ABCA, 27 March 2007, 2.

[7] Ibid.

[8] Ibid., 1.

[9] Major General J. P. Cantwell, AO, "Planning Guidance for Development of Cultural Understanding Capability in the Australian Army," dated 19 November 2007.

[10] Air, Land, Sea Application Center, "Cultural Impact on Tactical Operations," November 2006, (Final Coordination Draft), 3.

[11] Patrick Guinness, head of the School of Archaeology & Anthropology, Australian National University, Canberra, at a meeting with LtCol Daryl Campbell,

Australian liaison officer to the U.S Marine Corps, Quantico, VA, 16 August 2007.

[12] Montgomery McFate, "The Military Utility of Understanding Adversary Culture" *Joint Forces Quarterly* 38 (3d Qtr 2005): 43.

[13] Joint Operating Concept (DoD). "Irregular Warfare," Version 1.0, 11 September 2007, 2.

[14] Australian Army, "Adaptive Campaigning: The Land Force Integrated Response to Complex Warfighting," Version 4.15 (Canberra: Directorate Combat Development Future Land Warfare, 3 November 2006), 2.

[15] "Actors" are defined by Smith as all players in the system and involve friendly allied forces, foes, neutrals, civilians, NGOs, OGOs. Anybody capable of having an influence (either directly or indirectly) onto the system.

[16] "Fluxes" encompass all influences onto the discussed system; from National interests such as diplomatic, informational, military and economic through to the influence of a private soldier could have with a single shot. This influence should be viewed in a complex non-linear state and the reader should avoid traditional military 'cause and effect' linear though processes. Defined in James Moffat, *Complexity Theory and Network Centric Warfare* (Washington, DC: Command and Control Research Program Publication Series, 2003), 8-9 (http://www.dodccrp.org/files/Moffat_Complexity.pdf).

[17] Ibid., xi.

[18] The author has recommended task force and battle group level advisors as this has been the preponderance of force deployments in recent years. Only once has the division-sized headquarters with units deployed since Vietnam, and that was East Timor in 1999. All other deployments have been battalion or brigade. Higher headquarters have been a national command element or individual officers fulfilling positions on coalition deployments. If a situation arose such as a world war, or even another Vietnam commitment where the federal government deemed it necessary to conscript, then the human systems company would surge in numbers to complement.

Conclusion

Each of the chapters in this book offers a spectrum of cultural issues facing today's militaries as they prepare for war in the 21st century. As experienced military officers with direct operational experience in foreign cultures, the contributors to this book know intimately the practical "boots on the ground" issues that militaries face. Their essays all focus on the practical question of *how* to translate theoretical cultural concepts into concrete operational applications.

Major Jonathon P. Dunne examines the problems that arise when militaries simply transfer models that work in Western contexts to non-Western environments. Lieutenant Colonel Alejandro P. Briceno focuses on the challenges of managing an overwhelming amount of cultural information in a military environment where time and access to resources is limited. Major John E. Bilas' analysis illustrates the myriad practical difficulties that occur when Western militaries work alongside different cultural groups such as the Iraqi military. Challenging our military to redefine the problem from the perspective of the Afghan tribes, Major Randall S. Hoffman forces the reader to look at the issues from a new cultural lens. And both Major Robert T. Castro and Lieutenant Colonel Steven Brain confronts the dilemmas of transforming a military whose training continuum emphasizes reinforcing conventional skills into a military that is able to respond flexibly to complex irregular warfare environments.

It may seem that the many issues raised by the authors form part of a never-ending list of difficulties in incorporating culture into military operations. Yet on closer examination, the chapters in this book repeat several key common themes that affect the military's ability to plan and prepare for cultural factors. First, there is the issue of available time. Second, there are the limitations of the available skills and knowledge of the specific members in the military unit. Third, is the cultural reality of the actual operational situation in which the military unit finds itself—often a far cry from what they may have anticipated. And finally, there is the question of the organizational structure of the military unit itself.

Cultural Role of Time

Time is probably one of the greatest challenges in any military operation. Indeed, it is considered such a problem that Marines often cite the expression, "That four letter word: T-I-M-E." Three of the chapters in this book grapple with the question of how to deal with time when incorporating culture into operations. Lieutenant Colonel Briceno's model is a direct effort to help military leaders solve this problem. His essay seeks to help decision makers recognize which cultural factors are likely to be the most problematic in a particular area of operation. By reducing the cultural issues to several salient factors, the model enables leaders to focus their time most efficiently on key aspects of culture relevant to operations.

Major Bilas' article focuses on a reverse aspect of time: the cultural time of the Iraqi military. His analysis reveals that American notions of time fail to account for the social and cultural realities of the people in a foreign area of operations. By describing in great detail the specific practical issues of the Iraqi military, he explains how and why American expectations for developing a functioning Iraqi Army (IA) were unrealistic. By outlining the basic survival issues facing the IA—such as food, pay, medical care, and secure transportation—it becomes clear why the IA has had such difficulties in attracting and retaining personnel. His careful detailed analysis allows military planners to move beyond statements such as "Iraqis are lazy and don't understand time" to look at the practical and cultural issues that must be overcome to develop a successful efficient indigenous army.

Lieutenant Colonel Brain examines time from the perspective of training the Australian Army to be culturally effective. Given the constraints of time and the unpredictability of deployment destinations, he argues that it is impractical to train all of the Australian Army in the many cultures and languages of the hundreds of destinations where the army may deploy. His argument underscores a second key theme in the six chapters of this book: how to train and educate cultural skills across the spectrum of military personnel.

Military personnel

Recognizing that there is simply not enough time to develop an entire military that is culturally and linguistically competent, Lieutenant Colonel Brain offers a staged solution: one that focuses training on specific ranks and skill groups according to mission. He proposes accomplishing this by two approaches: first by developing the unique skill sets of specific military soldiers; and secondly by tailoring training to grade and rank.

First, as Lieutenant Colonel Brain's chapter emphasizes, not all military personnel have the necessary capacity to develop high level language and cultural skills. Nor are such skills necessary for personnel whose careers will rarely require direct interaction with local populations. He argues, therefore, for a tailored language and cultural instruction program that builds upon the existing skills and aptitudes of talented individual soldiers. By focusing training on a small group of capable soldiers, the Australian Army could expend its limited resources efficiently to create a cadre of skilled experts. Additionally, Brain suggests, the army could consider hiring outside experts such as anthropologists to serve as advisors; thus expanding its cultural skills without taking soldiers out of the military cycle for years of education.

Lieutenant Colonel Brain's second approach emphasizes that cultural skills development should match the career of Australian soldiers. He proposes a continuum beginning with cultural acknowledgment for entry level soldiers and progressing to what he terms, 'cultural cunning' for senior level officers.

Major Castro's chapter contemplates a similar progression for Marines in developing a culturally competent Corps. Comparing cultural skills training to leadership skills programs, he proposes that cultural education should be part of a lifelong progression over the career of a Marine: beginning with basic skills training upon entry and progressing to higher level educational and cognitive programs at the senior officer levels. Major Castro emphasizes the importance of professional military education (PME) in developing this progression. His model offers a tiered educational program that would tailor cultural education to rank, PME school, and assignment.

While time and the skills of military personnel are key issues in developing a culturally competent military, as many of the chapters illustrate, the ultimate test of cultural programs is found in the field. Successfully dealing with the cultural differences and people in one's area of operations is of course, the objective of any cultural program.

158 *Applications in Operational Culture*

Cultural Reality of the Area of Operations

As the chapters by Dunne, Briceno, Hoffman, and Bilas illustrate, perhaps the greatest challenge is understanding the cultural reality of the foreign peoples with whom the military must work. Major Dunne's chapter clearly demonstrates the hazards of transferring an American way of thinking about the world to a foreign operating environment. His analysis of the needs and concerns of the Iraqi people, from 2004 to present, illustrates the fallacy of assuming that a Western hierarchy of needs would mirror the values and priorities of the Iraqis. He argues most powerfully that in Iraq, an individual's self-actualization is not the highest goal, as Abraham Maslow posited. Dunne proposes, instead, that in Iraqi culture the success of the group (which he defines as group actualization) is valued much more highly.

Lieutenant Colonel Briceno's approach is also based on the assumption that there is a certain "cultural distance" between the values and ideals of Americans and other cultural groups. Indeed, his chapter offers a model that would help assess the degree that other groups differ from the U.S. on basic cultural concepts such as honor, use of land, gender etc. Using the case of Kuwait, he offers a visual analysis that enables decision makers to quickly grasp the most salient differences and incorporate them into planning and operations.

The chapter by Major Bilas provides an in-depth case study of the problems and challenges that occur when the concrete cultural realities of the AO are not taken into account. Following the challenges faced by the 2/7 BDE's Marine MiTT in Iraq, he explains how the initial training objectives of the MiTT were forced to shift due to the unanticipated cultural realities discovered on the ground. Although initially sent to provide intelligence support to the IA, Major Bilas actually spent much of his time on the ground trying to take care of basic logistics such as pay. His chapter emphasizes the importance of mission flexibility when working in a foreign AO.

Finally, Major Hoffman's chapter argues that in Afghanistan, the *people* are the center of gravity—a theme reiterated by Major Castro in the following chapter. According to Major Hoffman, in order to succeed in current operations in Afghanistan, the U.S. military must understand and successfully work with the Pashtun tribes. Citing the writings and perspectives of both the British and Russians in Afghanistan, he emphasizes the key role that tribal engagement has played in the successes and failures of our military predecessors in the country. Implicit in Major Hoffman's chapter is his emphasis on the ability of our own military to adapt its operations to function effectively in this foreign cultural environment.

Cultural Challenges of Our Own Military

As Major Hoffman's and Bilas' case studies illustrate, the ability of our own military to adapt to foreign cultures is as important to operational success as understanding the other culture. This issue is echoed throughout the book. Major Dunne's chapter clearly is a comment, not only on Iraqi culture, but on our own military's struggles to adapt our operations and strategy to the new cultural reality. While careful not to attribute the Anbar Awakening or other successes in Iraq specifically to the improved ability of the Marine Corps to apply operational culture principles, his paper clearly illustrates the successful shift that the Marine Corps has made from a more conventional Western approach to a culturally and tribally focused counterinsurgency strategy.

How to shift from a conventional military to one that is capable of succeeding in an irregular warfare environment is a key focus in Major Castro's chapter. He emphasizes the importance of using the professional military education system to develop, over the long term, a culturally competent Marine Corps. Facing the same issues for the Australian Army, Lieutenant Colonel Brain proposes a training continuum that would fit within the structure and organization of the existing Australian military culture rather than forcing the Australian Army to change fundamentally.

Preparing for the cultural challenges of warfare in the 21st Century will require important shifts in our current military structures: shifts in the way U.S. and allied militaries teach, train, organize, manage time and personnel and think about the nature of warfare and the role of local people in conflict. Each of the chapters in this book seeks to provide concrete examples and suggestions for incorporating culture into military operations. While much remains to be done, this book provides an important step forward: by seeking to learn from those who have been in the field, worked with foreign populations, and struggled with these questions on a day to day basis.

Ultimately, the true test of our cultural education and training programs will be whether or not we learn from our past mistakes. Within the pages of this book lie the records of some of those cultural lessons; lessons hard learned that hopefully, this time, will not be forgotten.

Appendix A

Iraqi Army Logistics Data, 2006-2007

	2nd BDE, 7th Iraqi Division BDE MiTT Personal Notes- 2006*	2nd BDE, 7th Iraq Division BDE MiTT Interview- 2007*	1st Iraqi Army Division, BDE MiTT MiTT AAR 2006*	1st Iraqi Army Division, BDE MiTT AAR 2007*	News Article June 2006*	News Article July 2007*
Logistics						
Quality of Life	Poor living conditions aboard the BDE and BN camps. Food contractor performed miserably. Soldiers would eat meat/produce only twice per week. No camp support, to include water removal. Total support required from the Regiment.	New Contractor, Sandi Group, provided life support for the BDE. Contract had American oversight but the quality of life improved for the soldiers. Fresh produce and meats began to arrive regularly. Road networks improved in Al Anbar because security was better. Numerous times, produce and meats was purchased in the local markets. MOD began allocating money to the BDE for local food purchases, which also showed signs of economy improvement.	The performance at Habbiniyah has been extremely poor since MOD took over control of awarding and paying for the contractor. Not enough supplies to support the BDE. The Army had to purchase food items in the open market with "out of pocket" money.	The contracting company, the Sandi Group, improved its performance. In January 2007, the contractor was providing 60% of the IA food and water. Spoilage, at times, was a problem, and they were not prepared for the "surge" of increased number of troops.	Lousy living conditions, bad food and failure to receive regular pay are the main reasons behind the exodus. In running at least several hundred soldiers a month, the officials said……Sometimes, they don't eat for two or three days at a time. Logistics has been the Iraqi army's primary problem here. The Iraqis complain most about the persistent problems with food and pay rather than bouts with combat and casualties.	The Iraqi military logistics system also is improving, Pine said, although he acknowledged it does experience occasional hiccups. "Logistics is probably the most complex thing any military force does, and so we're trying to really help them focus on the ability to do logistics," Pine said. Now paper-based, the Iraqi logistics system is being retooled to eventually incorporate a computer-run supply database patterned after one used by the U.S. Air Force, he said. he Iraqi army's logistics capabilities "have improved across the board," Pine said, noting efforts have been made to cut down on the amount of time it takes to approve supply requisitions from field units. In addition, the number of Iraqi army fuel requisitions that were filled by coalition sources dropped dramatically over the past few months, he said. Pine said, because the Iraqis' supply processes are improving.

Appendix A. Triangulated Data (Logistics) collected from various sources.

161

162   *Applications in Operational Culture*

Logistics	2nd BDE, 7th Iraqi Division BDE MiTT Personal Notes- 2006*	2nd BDE, 7th Iraq Division BDE MiTT Interview- 2007*	1st Iraqi Army Division, BDE MiTT MiTT AAR 2006*	1st Iraqi Army Division, BDE MiTT AAR 2007*	News Article June 2006*	News Article July 2007*
Fuel	All POL's (MOGAS, diesel) was provided by Coalition Forces aboard Camp Alasad. During 2006, there was no capability for fuel replenishment. Road networks and fuel tanker transportation was too dangerous to travel.	MOD began supplying fuel to the BDE and the BN's. The entire BDE was self sustained and required minimal assistance from Coalition Forces. Increase security allowed fuel trucks to travel freely throughout Al Anbar.	Supplemental fuel provided by Coalition Forces.	Within two months of January 2007, the BDE became totally independent because of the improved life support contract provided by the Sandi Group.		
Supply	Iraqi Soldiers had only one set of uniforms and one pair of boots. There was no supply system established during 2006. All supplies were supported by Coalition Forces.	The BDE's supply requisition improved. Able to conduct convoys to Div HQ (Ramadi) for supply replenishment and equipment. HUMVEE allocation increased through the Enhance Supplemental Utilization Program (ESUP) - temploan program of vehicles to the Army.	Army supply support was abysmal. No issue facility established for the BDE. Reports made to higher HQ but received no response. Morale greatly suffers. Cold weather gear not issued in 2005 and only 1/3 of equipment has arrived in 2006.	Logistical readiness improved to 98%. The BDE Commander and his staff had strong influence to supply and logistic readiness.		

Appendix A. Triangulated Data (Logistics) collected from various sources.

Appendix A 163

Logistics	2nd BDE, 7th Iraqi Division BDE MiTT Personal Notes-2006*	2nd BDE, 7th Iraq Division BDE MiTT Interview-2007*	1st Iraqi Army Division, BDE MiTT MiTT AAR 2006*	1st Iraqi Army Division, BDE MiTT AAR 2007*	News Article June 2006*	News Article July 2007*
Maintenance	National Maintenance Contract was established mid-2006. All preventative maintenance for Iraqi vehicles was done by the NMC, located adjacent to the DDE camp. The Iraqi BDE did no PM's on their vehicles, even though the MiTTs constantly pushed the Iraqis to do PM's.	Iraqis still do not do PM's. The NMC performed superbly during the year. However, the NMC does not have the spare parts block to repair the BDE's entire asset allocation. For example, Ford and Chevy pickup trucks have remained on a lot at the BDE HQ's for over one year.	Maintenance is non existent. Turn around is slow and usually takes months.	The Iraqi Army Maintenance Program (IAMP) improved the BDE's maintenance capability. The BDE has had success in sending vehicles to Habbiniyah for repair and maintenance*		

Appendix A. Triangulated Data (Logistics) collected from various sources.

164 Applications in Operational Culture

	2nd BDE, 7th Iraqi Division BDE MiTT Personal Notes-2006*	2nd BDE, 7th Iraq Division BDE MiTT Interview-2007*	1st Iraqi Army Division, BDE MiTT MiTT AAR 2006*	1st Iraqi Army Division, BDE MiTT AAR 2007*	News Article June 2006*	News Article July 2007*
Logistics						
Medical	Limited medical capability. Trained Combat Life Savers but the Army had no capability for any surgical care. All surgical care required Coalition Force support. The BDE had a Microbiologist as their Medical Officer. He was totally not qualified.	Medical capability improved slightly. All combat related injuries still required Coalition Force support. However, the BDE signed a MOU to the local hospital in Hit that all non-combat related injuries would be treated by the hospital.	No medical records exist for soldiers in the BDE. Preventative medicine practices do not exist. Medical readiness is not tracked.			
Camp Infra-structure	BDE was living in temporary quarters. Buildings were built by Combat Logistics Group. The buildings were wooden framed. They were heated/AC but was not built for permanent use.	The new camp was within three months of completion. The camp was nicely constructed for permanent infrastructure.				

Appendix A. Triangulated Data (Logistics) collected from various sources.

Appendix B

Iraqi Army Pay, 2006-2007

	2nd BDE, 7th Iraqi Division BDE MiTT Personal Notes-2006*	2nd BDE, 7th Iraqi Division BDE MiTT Interview-2007*	1st Iraqi Army Division, BDE MiTT AAR 2006*	1st Iraqi Army Division, BDE MiTT AAR 2007*	Battalion MiTT AAR comments December 2006
Pay					Four reasons Iraqis join the military: pay, vacation, quality of life. Pay was greatest motivator for individuals to remain on active duty.
Metrics	Over 400 "ghost soldiers" on rolls. 100% MiTT involvement to ensure rosters are clean. Required a lot of MiTT presence at MOD to reconcile rosters. Average 97% of BDE was paid monthly.	Process improved dramatically in 2007. 98.5% of BDE paid monthly. Ghost soldiers was at 40 during in December 2007	25-30 Soldiers were never paid on time. 80% of RIbE paid.	Early 2007, 700 ghost soldiers were on the rolls. By end of 2007, it reduced to 17. No pay dues reduced during 2007.	
Process	Complex process of getting check signed monthly. The check process required numerous stamps and signatures. Not efficient process. Always wondered of the significance of the MOD workers. I attributed to their culture.	Cashing check improved to central location in Green Zone (Bank within MOD). Eliminated convoy support to local bank and was more secure. Still required many stamps and signatures.	New joined the BDE, it was impossible to for the Battalions/Brigade/Division to accurately and quickly process these soldiers into the MOD system.	Required MiTT involvement to address issues at MOD level. MiTTs would have "ticket in" to MOD to address issues face to face with MOD personnel.	

165

166 *Applications in Operational Culture*

	2nd BDE, 7th Iraqi Division BDE MiTT Personal Notes–2006*	2nd BDE, 7th Iraq Division BDE MiTT Interview–2007*	1st Iraqi Army Division, BDE MiTT AAR 2006*	1st Iraqi Army Division, BDE MiTT AAR 2007*	Battalion MiTT AAR comments December 2006
Pay					
Process cont'd	Pay process was complex and was often delayed. Required helo lift to Baghdad, meetings with MOD to sign for check, coordinate convoy to local bank to cash check, drive back to Green Zone, fly back to Al Asad, convoy to Bns to pay soldiers.	Payment to soldiers remained consistent at around the 5th of each month. Reconciliation became easier by eliminating trip to Division HQ.	Delay in paying soldiers resulted in soldiers quitting, which had a significant impact on operational readiness.	MiTTs and G-1's would send Arabic/English report to MOD for corrective action.	
	Security situation in Baghdad delayed cashing of check on three occasions, further complicating logistic movement and coordination.	Data base being established in Al Anbar–(HRIMS) Human Resource Information Management System. Similar to Marine Corps 3270. Expected to eliminate corruption and improve reconciliation process.		Recommend a "diary" system to effect corrective actions.	

Appendix A. Triangulated Data (Pay) collected from various sources.

	2nd BDE, 7th Iraqi Division BDE MiTT Personal Notes–2006*	2nd BDE, 7th Iraq Division BDE MiTT Interview–2007*	1st Iraqi Army Division, BDE MiTT MiTT AAR 2006*	1st Iraqi Army Division, BDE MiTT AAR 2007*	Battalion MiTT AAR comments December 2006
Pay					
Process cont'd	MiTTs closely watched pay officers pay the soldiers. Properly identifying soldiers was difficult for the MiTTs to determine if he was the actual person on the pay roster.	Pay officers were replaced and new Iraqi BDE commander instilled discipline into the BDE. MiTTs gained trust in the pay officers. In addition, all soldiers are BAT'd and have official MOD Iraqi Army ID cards. Soldiers must present their ID cards upon receiving their payment.			

Appendix A. Triangulated Data (Pay) collected from various sources.

168 Applications in Operational Culture

	2nd BDE, 7th Iraqi Division BDE MiTT Personal Notes-2006*	2nd BDE, 7th Iraq Division BDE MiTT Interview-2007*	1st Iraqi Army Division, BDE MiTT MiTT AAR 2006*	1st Iraqi Army Division, BDE MiTT AAR 2007*	Battalion MiTT AAR comments December 2006
Pay					
Process cont'd	Same nut roll to reconcile pay rosters and return cash back to Baghdad. Flight coordination and getting MOD to respond to fixing the pay rosters was challenging.	Money is returned directly to MOD within Green zone. No convoy support required.			
	Only one central bank system. Had opportunity between May thru July to use local bank in Hadithah, but poor security halted process. The bank manager feared for his life.	MOD now exploring local branch banks in Fallujah, Ramadi, and Hit to support Army payroll.			
Process cont'd	Slow responsiveness of MOD to not fix rosters led to soldiers quitting the Army.	New HRIMS will eliminate paper trail. Pay rosters will be updated automatically.			

Appendix B 169

Pay	2nd BDE, 7th Iraqi Division BDE MiTT Personal Notes-2006*	2nd BDE, 7th Iraq Division BDE MiTT Interview-2007*	1st Iraqi Army Division, BDE MiTT MiTT AAR 2006*	1st Iraqi Army Division, BDE MiTT AAR 2007*	Battalion MiTT AAR comments December 2006
Soldier's pay levels	Soldiers complained that they did not make enough money. Always complained about not receiving hazard pay. Soldiers received 477,000 dinar per month/ $320.00 per month	Soldiers received $160,000 increase in pay or $450 per month.			
Soldier's pay levels cont'd	Soldiers received cash payment. Money was brought back to their home to pay their families.	MOD is exploring a direct deposit system for Army soldiers. Direct deposit is new to the Iraqi people but is utilized with MOD employees at this time.			
		Increased pay helped recruiting effort in Al Anbar in 2007. Increase pay also improved soldier's quality of life.			

Appendix A. Triangulated Data (Pay) collected from various sources.

Appendix C

Growth of the Iraqi Army

Date					
May 2003	7,000 - 9,000	N/A	0	N/A	7,000 – 9,000
June	N/A	N/A	0	N/A	N/A
July	30,000	N/A	0	N/A	30,000
August	34,000	670	0	2,500	37,170
September	37,000	2,500	0	4,700	44,200
October	55,000	4,700	700	6,400	66,800
November	68,800	12,700	900	12,400	94,800
December	71,600	15,200	400	12,900	99,600
January 2004	66,900	19,800	1,100	21,000	108,800
February	77,100	27,900	2,000	18,000	125,000
March	75,000	33,560	3,005	23,426	134,991
April	80,016	23,123	2,367	18,747	124,253
May	90,803	24,873	3,939	16,097	135,712
June	83,789	36,229	7,116	18,183	145,317
July	31,300	36,229	7,700	19,859	95,088
August	32,942	37,925	6,288	14,313	91,468
September	40,152	36,496	7,747	14,313	98,708
October	44,728	41,261	6,861	18,148	110,998
November	49,455	43,445	6,013	14,593	113,506
December	53,571	40,115	14,500	14,267	118,009
January 2005	58,964	36,827	14,796	14,786	125,373
February	82,072 "trained and equipped"	"operational"	59,689	N/A	141,761 Trained and Effective: General Myers: 40,000 Senator Biden: 4,000 - 18,000
March	84,327		67,584	N/A	151,618 Trained and Effective: Lt. Gen Petraeus: 50,000 "off-the-cuff"
April	86,982		72,511	N/A	159,493
May	91,256		76,971	N/A	168,227
June	92,883		75,791	N/A	168,674
July	94,800		79,100	N/A	173,900 26,000 in Army in level I and II
August	101,000		81,900	N/A	182,900
September	104,300		87,800	N/A	192,100 - 30,000 in Army in level I and II 56
October[37]	111,000		100,000	N/A	211,000 -32,000 in level I and II 58
November	112,000		102,000	N/A	214,000
December	118,000		105,700	N/A	223,700
January 2006	120,400		106,900	N/A	227,300
February	123,600		108,500	N/A	232,100 -46,000 MOD forces and 8,000 MOI forces in Level I and II 59
March	134,800		115,700	N/A	250,500
April	138,700		115,000	N/A	253,700
May	145,500		117,900	N/A	265,600
June	148,500		116,100	N/A	264,600
July	154,500		115,100	N/A	269,600
August	167,900		130,100	N/A	298,000
September	176,200		131,600	N/A	307,800
October	180,800		131,600	N/A	312,400
November	188,300		134,700	N/A	323,000
December	188,300		134,700	N/A	323,000
January 2007	188,300		134,700	N/A	323,000
February	188,260		134,920	N/A	323,180
March	193,300		136,500	N/A	329,800
April	193,300		139,800	N/A	333,100
May	194,200		154,500	N/A	348,700
June	194,200		158,900	N/A	353,100
July	194,200		158,900	N/A	353,100
August	194,200		165,500	N/A	359,700
September	194,200		165,500	N/A	359,700
October	194,200		165,500	N/A	359,700
November	238,089		191,541	N/A	429,630
December	210,529		194,233	31,431	439,678

Source: Iraq Index, Tracking Variables of Eeconstruction and Security in Post Saddam Iraq, Brookings Institution, 21 December 2007, 15.

Appendix D

Case Examples of Culture and Operations Iraq

Case One

In al-Anbar Province in 2005, 3d Battalion, 8th Marines (3/8), was experiencing a high volume of VBIEDs in their Area of Operation Responsibility (AOR). During an investigation being conducted by Explosive Ordinance Disposal (EOD), a Marine providing security from 3/8 suggested to the investigator to look for the person who did the welding to the vehicle. The Marine went on to explain that "every welder has a distinct weld, just like a fingerprint," so if you find the welder, you will have your man. Within one week of the decision to find the welder, who made the vehicle, the battalion was successful in finding the welder and making the arrest. The reason this happened so quickly was due to the battalion S-3 collecting information back from the Marines after their patrols. This information was used to develop a cultural map of the AOR listing where people lived, business locations, and most importantly, what type of businesses.

Case Two

From my unit's experiences in the Fallujah in 2005, while conducting dismounted foot patrols in the Jolan District of the city, it became very apparent that we were in a rapidly changing situation. The street was vacant, with the women in the market areas missing and men huddled inside rather than standing outside. The patrol took a defensive approach to the situation realized that there was an IED was in place on the corner of a nearby intersection. Due to our understanding of normal market activity in the area, we were able read the actions of the local population as danger signs, which gave us fair warning of the imminent threat.

www.ingramcontent.com/pod-product-compliance
Lightning Source LLC
Chambersburg PA
CBHW070756100426
42742CB00012B/2149